Sample Preparation Techniques for Chemical Analysis

Edited by Massoud Kaykhaii

Published in London, United Kingdom

IntechOpen

Supporting open minds since 2005

Sample Preparation Techniques for Chemical Analysis
http://dx.doi.org/10.5772/intechopen.96639
Edited by Massoud Kaykhaii

Contributors
E. Celeste Welch, Massoud Kaykhaii, Mona Sargazi, Sayyed Hossein Hashemi, J. Paul Robinson, Harekrishna Panigrahi, Smrutirekha Mishra, Suraj Kumar Tripathy, Anubhav Tripathi

First published in London, United Kingdom, 2021 by IntechOpen
IntechOpen is the global imprint of INTECHOPEN LIMITED, registered in England and Wales, registration number: 11086078, 5 Princes Gate Court, London, SW7 2QJ, United Kingdom
Printed in Croatia

British Library Cataloguing-in-Publication Data
A catalogue record for this book is available from the British Library

Additional hard and PDF copies can be obtained from orders@intechopen.com

Sample Preparation Techniques for Chemical Analysis
Edited by Massoud Kaykhaii
p. cm.
Print ISBN 978-1-83969-214-7
Online ISBN 978-1-83969-215-4
eBook (PDF) ISBN 978-1-83969-216-1

For EU product safety concerns:
IN TECH d.o.o., Prolaz Marije Krucifikse Kozulić 3, 51000 Rijeka, Croatia, info@intechopen.com or visit our website at intechopen.com.

Meet the editor

Massoud Kaykhaii has been a Professor of Analytical Chemistry at the University of Sistan and Baluchestan (USB), Zahedan, Iran, since 1989. His research work focuses on modern sample preparation techniques including miniaturized solid and liquid phase microextraction and stir bar sorptive extraction. He has written 150 research articles, 5 books, and 22 national standard procedures of analysis. He has three patents to his credit. He has presented at 150 seminars/conferences and (co)supervised 106 MSc and Ph.D. theses. He is a member of the editorial advisory board of ninety-eight journals and acted as secretary of three national mirror committees of ISO/TC. Professor Kaykhaii was recognized as the best researcher at USB and as being in the top 1 percent of reviewers in chemistry by Publons.

Contents

Preface

Sampling and sample preparation are major parts of every chemical analysis. These steps are the most time-consuming and challenging of any analysis procedure. In addition, most errors are likely to occur at these steps. This is why the sample preparation step is known as the "bottleneck" of any analysis; therefore, the overall success of analysis depends on advances in sample preparation methods.

For many years, sample preparation techniques were based on low-tech approaches including non-automated and labor techniques such as dilution, filtration, deproteinization, derivatization, centrifugation, purification, and extractions such as liquid-liquid extraction and liquid-solid extraction (i.e., Soxhlet extractor). These techniques are hazardous to both the environment and operator and require a high volume of samples and extracting organic solvent. This book is divided into three sections that discuss a variety of sample preparation techniques and their applications.

Chapter 1, "Introductory Chapter: Evolution of Sample Preparation", reviews the development of sample preparation techniques away from hazardous methods toward modern and more environmentally friendly approaches. Evolving rules of green analytical chemistry have forced analytical chemists to find new, "greener" sample preparation and extraction techniques. This has resulted in two major advances in sample preparation in the past three decades. First is the introduction of solid-phase extraction (SPE), and the second is the invention of solid-phase microextraction (SPME). Both approaches were themselves revolutionized with the introduction of nanomaterials as sorbents. While SPE is an exhaustive method of separation, SPME is a non-exhaustive method of extraction based on equilibrium. SPME tries to address solvent-free needs. Both SPE and SPME can be fully automated, and by changing the type of their solid or pseudo-solid extracting phase, they can be used for many varieties of analytical needs. Chapter 2, "Modern Sample Preparation Techniques: A Brief Introduction," details the history and different modes of extraction techniques, derived mainly from SPE and SPME. Together, Chapters 1 and 2 introduce readers to the basic concepts in sampling and describe in detail the major techniques. They also discuss automation, miniaturization, advantages, and some applications of the presented techniques.

The second section of the book presents the latest advances in sample preparation techniques for surface spectroscopy analysis. Both theoretical aspects of sample preparation and practical recipes for successful surface analysis are discussed.

Finally, the third section of the book discusses sample preparation for particle, tissue, and cellular separation in two chapters: "Preparation of Tissues and Heterogeneous Cellular Samples for Single-Cell Analysis" and "Particle and Cell Separation."

Although this book does not provide a comprehensive review of all techniques of sample preparation, it does present the latest information to keep readers abreast of new areas of research in this field.

I am grateful to all the contributing authors for their time and efforts in preparing the chapters of this book.

Massoud Kaykhaii
Chemistry Department,
University of Sistan and Baluchestan,
Zahedan, Iran

Section 1

Advances in Sample Preparation

Chapter 1

Introductory Chapter: Evolution of Sample Preparation

Massoud Kaykhaii

1. Importance of sample preparation in chemical analysis

The start of chemistry as a science can be dated back to the time that the textbook "Traité élémentaire de chimie" was written by French chemist Antoine-Laurent Lavoisier in 1789. This book can be regarded as the beginning of entry on measurement in chemistry. Lavoisier wrote: "When *16 ounces* of alcohol are burnt in an apparatus properly adapted for collecting all the water disengaged during the combustion, we obtain from *17 to 18 ounces* of water" [1], which clearly shows that a "gravimetric analysis" was performed during his experiments. In today's modern chemical analysis, still balances are the basis of all measurements. An analytical method is carried out to help us gather "analytical data" to solve problems. From a general point of view, these data can be either qualitative or quantitative. Qualitative analysis tries to find an answer to the question yes/no to the presence of a particular analyte in a sample, while quantitative analysis provides an answer to the basic question of "how much" of an analyte of interest presents in a sample. There is another term, semiqualitative analysis, which concerns whether the analyte is below or above a certain concentration. Each chemical analysis needs to be performed in several consecutive steps; sampling, sample preparation (including dilution, filtration, deproteinization, derivatization, centrifugation, purification, isolation and separation, extraction and preconcentration), quantitative analysis (i.e., introducing (the extract of) sample into an analytical instrument), statistical evaluation of the obtained data (i.e., instrument's output), and making decision. Extraction is the major part of sample preparation and separation and can be considered as the partition of a solute between two immiscible solvents. One solvent contains samples with analytes and the second one is the extraction phase. It should be noted that before completion of each step, the next step cannot be started. Moreover, the raise of any error in each step means that the chemical analysis process needs to be repeated from the same step. It is well known that what causes most errors in chemical analysis is the sample preparation step (**Figure 1**) [2]. This step is also the most time-consuming and also is highly laborious [3]. Even having the highest expertise and best laboratory chemical, devices, and glassware, total systematic and random errors for sampling and sample preparation can occur from a few percentages to several orders of magnitude. As a result, any progress, improvement, or optimization in this "bottle neck" step of a chemical analysis process greatly boosts performance of the overall analysis, including precision, accuracy, and rapidity of the method. On the other hand, while analytical instrumentation in the recent years has continued to benefit from improvements such as tiny size, fastness, high resolution, low cost of capital investment, maintenance, and operation, still their limit of detection is not enough to be able to directly analyze a sample with ultra-trace amounts of a particular analyte. Besides, interferences from the matrix of the sample and its incompatibility (due to the dirtiness of real samples) with an

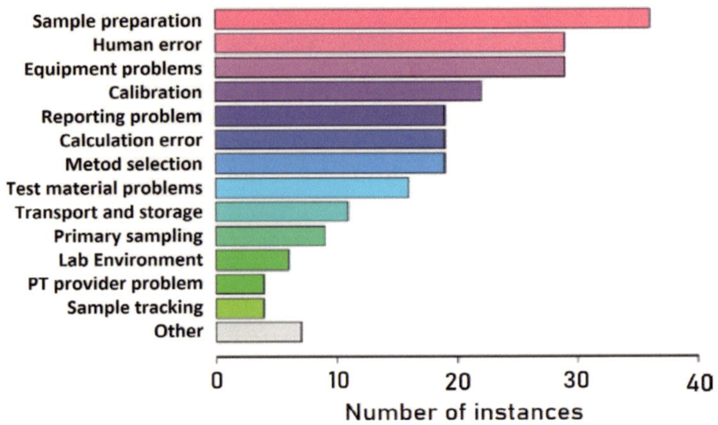

Figure 1.
Causes of error in chemical analysis (copyright 2013, Royal Society of Chemistry. Reprinted from Ref. [2] with permission).

analytical instrument make a cleanup and sample preparation step indispensable. Even with higher concentration of the analyte, this step is necessary for protecting instruments, for example, to increase the life of chromatographic columns.

2. Exhaustive and nonexhaustive extraction

Traditionally, sample preparation consisted of sample dissolution, purification, and extraction that were carried out with liquid-liquid extraction (LLE). For many

Figure 2.
"They have the very best instrumentation, for example to determine the composition of drinking water at the sub-parts per trillion level with triple quadrupole mass specs and gas chromatography. But do you know what their sample preparation is? They take one liter of water, add 100 ml of dichloromethane, shake it for half an hour. Then they put the dichloromethane in a vial, evaporate it down to 0.5 ml—don't worry, it's just going into the air—and then they inject. We need to do something about that..." Courtesy of Analytical Scientist, 27/7/2014.

years, classical extraction methods as sample preparation tools were in use. For example, Soxhlet extractor invented in 1879 by Franz von Soxhlet still is widely in use.

Depending on the receiving phase in contact with the sample solvent, extraction can be either sorbent-based (in case of solid extracting phase) or solvent-based (in case of liquid extracting phase). Through the years, both methods are applied in many formats and versions and are simple and effective. They are classified as exhaustive extraction techniques, which mean a significant amount of the target analyte is extracted to the receiving phase during extraction. The disadvantages with liquid-liquid extraction include the use of large volumes of organic solvent (**Figure 2**), cumbersome glassware, and high cost. Furthermore, LLE often creates emulsions with aqueous samples that are difficult to extract, and it is not easily automated. These limitations were overcome with solid-phase extraction (SPE) invention in the mid-1970s. Besides reduction of volume of organic solvents, SPE showed advantages such as miniaturization of the device, multiclass compound extraction and automation, and/or high-throughput determination [4]. Solvents in the order of 100 ml, which are needed in LLE, are reduced to a few milliliters in SPE. Moreover, normal phase, reversed phase, ion exchange, mixed mode, and selective sorbents (such as molecularly imprinted polymers) are now available commercially to cover any type of analytical extraction.

3. Miniaturized extraction

Solid-phase microextraction (SPME) is a simple and efficient, solventless sample preparation method, invented by J. Pawliszyn in 1989 as an attractive choice primarily applied for volatile organic compounds extraction applications [5]. This method opened a new door to the sample preparation technique, known as microextraction methods nowadays. Microextraction methods are based on equilibrium between sample solution (or its headspace) and extraction phase, which has much smaller volume than the sample itself. While there were major drawbacks such as fragility of the SPME fibers, its majority of applications was limited to gas chromatography; after commercialization, it became the most popular microextraction. In 1999, J. Namiesnik proposed the term green analytical chemistry (GAC) [6], and then 12 principles of GAC were proposed by A. Gałuszka et al. as: 1. direct analytical techniques, 2. minimize sample size and number of samples, 3. in situ measurements, 4. integration of analytical processes and operations to reduce energy and reagents usage, 5. automation and miniaturization of methods, 6. avoid derivatization, 7. avoid waste generation and provide appropriate waste management, 8. multi-analyte or multi-parameter methods, 9. minimize the use of energy, 10. preferable use of reagents obtained from renewable sources, 11. elimination of toxic reagents, 12. increase of the safety of the operator [7]. All these rules are compliant in microextraction methods, which resulted in these newly introduced techniques becoming at the center of attention of analytical chemistry researchers (**Figure 3**) [8]. These techniques were also greener sample preparation methods in terms of waste generation and energy consumption. From the point of view of the above 12 principles, methods of in-field analysis can be considered as the foremost green techniques, which were in use before microextraction. These techniques became even more efficient and greener by the introduction of smart phones into analytical chemistry as the detection and data analysis/transmission devices [9]. Specialized sample preparations, such as self-assembly of analytes on nanoparticles for surface enhancement, have also evolved in parallel. Another aspect that makes microextraction methods more in demand is because modern analytical chemistry tends to prepare methods that are more around interdisciplinary research areas in biotechnology, environmental science, and materials science, etc.

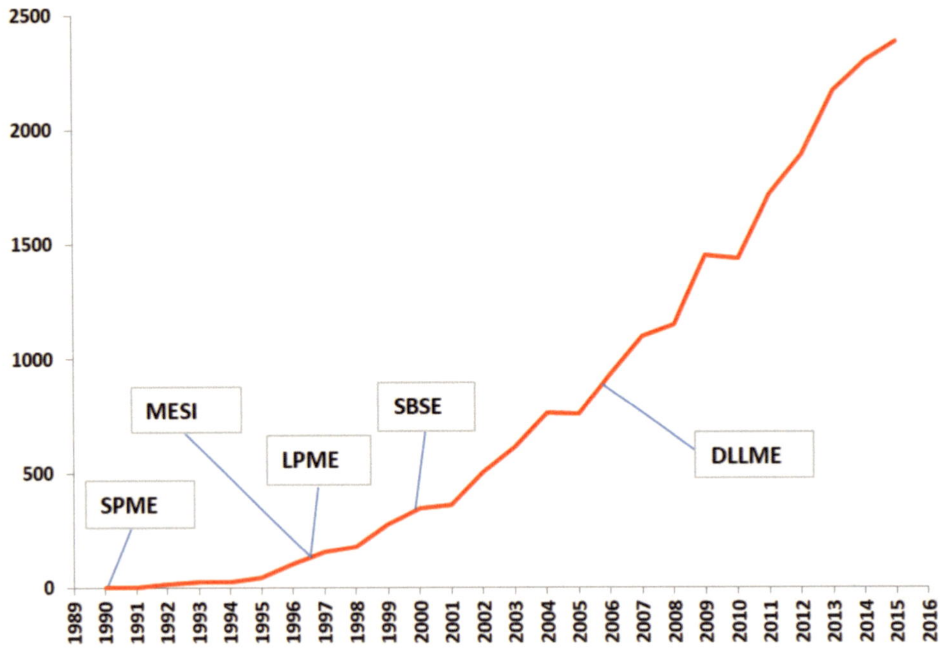

Figure 3.
Number of papers published since the introduction of SPME on applications of microextraction to chemical analysis. SPME, solid-phase microextraction; MESI, membrane extraction with a sorbent interface; LPME, liquid-phase microextraction; SBSE, stir bar sorptive extraction; DLLME, dispersive liquid-liquid microextraction (copyright 2017 Taylor & Francis Group, LLC. Reprinted from Ref. [8] with permission).

Analytical microextraction can be defined as a sample preparation step using volumes in the microliter or nanoliter range of extracting phase (solid, semisolid polymeric, or liquid material) [4]. They include 12 principles of GAC for sample treatment, especially in terms of miniaturization, in-field analysis, automation, and rapidity. These techniques are equilibrium-based extraction methods, which mean that the analyst may not know when equilibrium is reached, and the equilibrium distribution may necessitate multiple extractions [10]; this is the main drawback of these techniques. Besides SPME, analytical chemists invented a wide range of other microextraction techniques, including: single-drop microextraction, microextraction in packed syringe, membrane extraction with a sorbent interface, liquid-phase microextraction, dispersive liquid-liquid microextraction, accelerated solvent extraction, vortex-assisted microextraction, dispersive solid-phase extraction, pressurized liquid extraction, fabric phase sorptive extraction, micro solid-phase extraction, microwave-assisted solvent extraction, stir bar sorptive extraction, quick, easy, cheap, effective, rugged, and safe extraction (QuEChERS), magnetic solid-phase extraction, salt saturated single-drop microextraction, micro-cloud point extraction, matrix solid-phase dispersion, pipette tip micro solid-phase extraction, and many more.

Unluckily, in many universities and analytical chemistry textbooks, importance and steps of sampling and sample preparation are not mentioning during analytical courses. It is always to be remembered that because of the importance of these steps in the generation of precise and accurate qualitative, semiquantitative, and quantitative analytical data, such skills are subject of great importance to be taught to the students. In terms of routine application of miniaturized microextraction methods in standard operating procedures, microextraction requires more time, but once the procedure is optimized, these methods are more convenient, greener, and more economical in comparison to traditional exhaustive extraction methods.

Author details

Massoud Kaykhaii[1,2]

1 Department of Chemistry, Faculty of Sciences, University of Sistan and Baluchestan, Zahedan, Iran

2 Department of Process Engineering and Chemical Technology, Faculty of Chemistry, Gdansk University of Technology, Gdansk, Poland

*Address all correspondence to: kaykhaii@gmail.com

References

[1] Michael W, Needham P, Hendry R. Philosophy of chemistry. In: Zalta EN, editor. The Stanford Encyclopaedia of Philosophy. Spring 2019 ed. Available from: https://plato.stanford.edu/archives/spr2019/entries/chemistry

[2] Analytical Methods Committee. What causes most errors in chemical analysis? AMCTB No 56, Analytical Methods. 2013; 5, 2914-2915, DOI: 10.1039/C3AY90035E

[3] Picó Y. Recent advances in sample preparation for pesticide analysis. In: Pawliszyn J, editor. Comprehensive Sampling and Sample Preparation. 1st ed. Academic Press: Amsterdam; 2012. pp. 569-590. DOI: 10.1016/B978-0-12-381373-2.00101-0

[4] Valcárcel M, Cárdenas S, Lucena R. Analytical Microextraction Techniques. 1st ed. Sharjah: Bentham Science Publishers; 2017. pp. 4-5

[5] Belardi RG, Pawliszyn J. The application of chemically modified fused silica fibers in the extraction of organics from water matrix samples and their rapid transfer to capillary columns. Water Pollution Research Journal of Canada. 1989;**24**:179-191. DOI: 10.2166/wqrj.1989.010

[6] Namiesnik J. Pro-ecological education. Environmental Science and Pollution Research. 1999;**6**:243-244. DOI: 10.1007/BF02987339

[7] Gałuszka A, Migaszewski Z, Namiesnik J. The 12 principles of green analytical chemistry and the SIGNIFICANCE mnemonic of green analytical practices. Trends in Analytical Chemistry. 2013;**50**:78-84. DOI: 10.1016/j.trac.2013.04.010

[8] Kaykhaii M, Linford M. Application of microextraction techniques including SPME and MESI to the thermal degradation of polymers: A review. Critical Reviews in Analytical Chemistry. 2017;**47**:172-186. DOI: 10.1080/10408347.2016.1242062

[9] Koohkan R, Kaykhaii M, Sasani M, Paull B. Fabrication of a smartphone-based spectrophotometer and its application in monitoring concentrations of organic dyes. ACS Omega. 2020;**5**:31450-31455. DOI: 10.1021/acsomega.0c05123

[10] Wells MJM. Handling large volume samples: applications of SPE to environmental matrices. In: Simpson NKJ, editor. Solid-Phase Extraction: Principles, Techniques, and Application. 1st ed. Boca Raton, FL, USA: CRC Press; 2000, Chapter 14

Chapter 2

Modern Sample Preparation Techniques: A Brief Introduction

Mona Sargazi, Sayyed Hossein Hashemi and Massoud Kaykhaii

Abstract

Due to fast growth in microprocessors, analytical instrumentations in spectroscopy, chromatography, microscopy, sensors and microdevices have been subjected to significant developments. Despite these advances, a sample preparation step is indispensable before instrumental analysis. Main reasons are low sensitivity of the instruments, matrix interferences and incompatibility of the sample with the analytical device. Most of the time spent and most of the errors occurring during a chemical analysis is on sample preparation step. As a result, any improvements in this essential process will have a significant effect on shortening the analysis time and its precision and accuracy and lowering the cost. This introductory chapter intends to draw the readers' attention to the importance of sample preparation, the procedures of sampling and the source of errors that occur in the course of sampling. The chapter then continues with a heading on sample preparation techniques, including exhaustive and non-exhaustive methods of extraction. Microwave, sonication and membrane-based extraction techniques are more emphasized as exhaustive methods and under a new title, miniaturized methods are discussed. Automation, on-line compatibility and simplification is an important aspect of any sample preparation and extraction which is discussed at the end of this chapter.

Keywords: sample preparation, matrix interferences, microextraction, automation

1. Introduction

Over the last two decades, efforts in sample preparation have been done to eliminate organic solvents and perform rapid analysis of combinatorial chemistry and biological samples, which require a high level of automation. These new developments in sample preparation, as a result of miniaturization of the extraction process, were a reason to extend new solvent-free approaches. Also, a fundamental understanding of extraction principles has been very important in the development of novel approaches, which results in new trends in sample preparation, such as microextraction, miniaturization, and integration of the sampling and separation. The sampling and sample preparation process, which is similar to engineering approaches on a smaller scale nevertheless is different from those related to chromatographic separations or other traditional disciplines of analytical chemistry, consists of extraction of components of interest from the sample matrix to an extracting phase. So, the procedure of extraction has a variety of selectivity, speed, and convenience as a result of the approach and condition used and geometric configurations of the extraction phase [1]. Modernization of analytical methods and instrumentation made it possible to measure smaller concentrations of even the most complex molecules and species in complicated matrices. In addition, improved measurement techniques and tools allow, or often require, the use of smaller analytical test portions to determine analyte concentrations. Using small test portions face with difficulty in achieving representativeness of the population, especially in the analysis of trace components, since the quality of any analytical result, depends on sample representativeness and integrity. Similar to all measurements, which are accompanied by errors, some typical errors often occur during sample collection prior to trace analysis [2] but many sources of error in an analysis can be controlled through the use of blanks, standards, or reference samples. Neither blank nor standard can repair the damage caused by an invalid sample [1]. Errors are an integral part of all measurements, which cannot be eliminated completely but it is possible to estimate the magnitude and nature of them in order to validate results [3].

In this chapter, an attempt has been made to define the principle of sample preparation and its importance prior to chemical analysis. The following sections are dedicated to different methods of sampling and the errors that occur during sampling and focus on the different techniques in the science of sample preparation. This chapter is not intended to provide a comprehensive review of the topic of sample preparation, but it is helpful for learning more about it.

2. What is sampling and sample preparation?

Successful quantitative analysis starts with the sample collection, which is a first operation in an analytical procedure. The sample in sample collection should be representative of the bulk material and various sample collection techniques can be employed depending on the nature of the matrix. For example, in the field of environmental samples, such as soil, air, and water, consideration of representative sampling right at the point of collection is of great importance because it needs to consider the intrinsic heterogeneity of most materials. These considerations are highly important especially in performing analysis of trace and ultra-trace components, which require sample-specific strategies to obtain a clear and unbiased overview. After sample collection, it needs to be decided whether the entire sample or a portion of the sample requires to be analyzed [4]. It should be noted that sampling depends on the location, depth, and time of the year, which affects the concentration of the sample. Once the sample is ready for analysis, sample preparation is the next step [3]. Sample preparation is a crucial step of the analysis process with the ability to detect an analyte at a level that is appropriate for detection by the instrument. Matrix interferences in this step can interfere with the detection and measurement of analytes in a way that often requires to complete separation of the sample matrix from the analyte(s) of interest. So, the nature of the analyte and the matrix determine the choice of sample preparation. **Figure 1** shows various sample preparation steps that may be employed in sample preparation in which most analysts use one to four steps for sample preparation, although, in some cases, more than seven steps may be used [4].

Nowadays, efforts have been done to hyphenate sampling and sample preparation steps but there are challenges such as multistep procedures involving organic solvents. So, it is difficult to develop a method to integrate sampling and sample preparation with separation methods, for the purpose of automation. Therefore, over 80% of analysis time is currently spent on sampling and sample preparation steps for complex samples [1, 5].

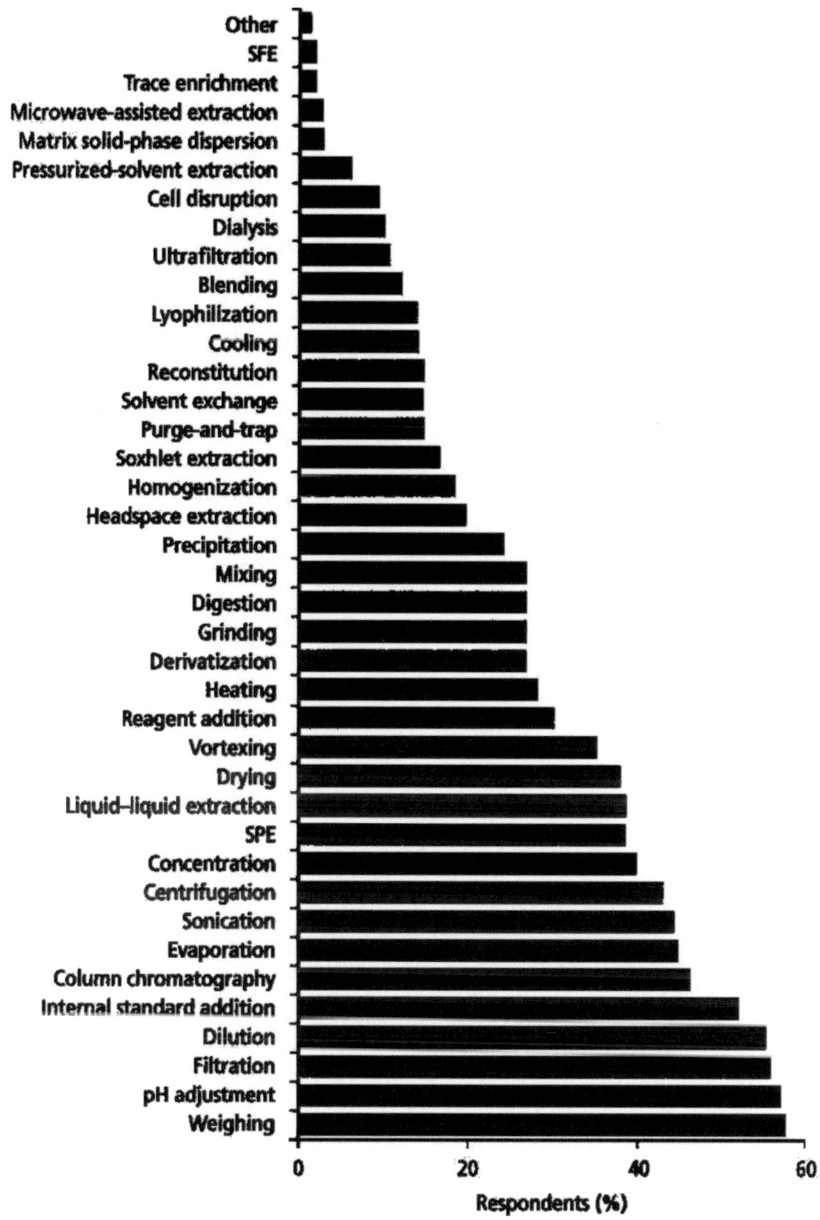

Figure 1.
Sample preparation procedures commonly used [4].

3. Importance of sampling and sample preparation

One of the most important parts of the analytical process is sampling, which is highly dependent on the properties of the analyte and the nature of the sample. Obtaining correct and informative results from the analytical procedure is a main purpose of this step. Inappropriate sample collection causes irreparable impairment that cannot be compensated even by quality assurance measures. Sample contamination during sample collection is one of the main sources of acquiring invalid data [6]. The decision of where to collect samples, which represents the whole sample, how to characterize the problems, and choosing a right method of obtaining the right amount of samples, are the critical issues in the sampling step. The purity of the sample should be ensured before taking a measurement to obtain the optimum results when using any instrument, irrespective of the technology. For this reason, sample preparation often involves a cleanup step for "dirty" samples besides extraction procedure. Sample preparation also includes all treatments to decompose the structure of the matrix in order to perform the fractionation, isolation, and enrichment of the proposed analytes. These treatments, which adapt the tested analytes with the detector to enhance the sensitivity of the detector, are also considered part of the sample preparation protocol. Emerging some challenging problems with the sample matrix led analytical chemists to turn to non-traditional technologies, which meet the need for solvent-free approaches, automation, and miniaturization. These developments reach some benefits such as *in situ* analysis of sample, which consequently reduces the time and errors and thus leads to more accurate, precise, and faster data. In parallel with the development of new technologies, the fundamental of extraction principles has advanced, which is in high importance in the development of novel sample preparation trends. In the case of complex samples, the analytical procedure consists of several steps including sampling, sample preparation, separation, quantification, statistical evaluation, and decision making. These analytical steps follow one after another, and the slowest step determines the whole speed of the analytical process. Since direct analysis cannot produce real results because of interferences of a matrix of the sample that causes low sensitivity, therefore, modification of the sample to obtain better analytical results is very important [7, 8].

4. Sampling methods

The aim of sampling methods is to choose a "sample" that represents the population. Attempts in sampling methods are ensuring about an equal chance of each part of the population to be selected for analysis, which requires a random element in the sampling strategy. These strategies include simple random sampling, systematic grid sampling, stratified, cluster, and two-stage sampling.

In simple random sampling, the population is divided into a set of units and a sample is selected unit by unit with an equal probability of selection for each unit at each draw. In systematic grid sampling, samples are collected from grid areas, which are as a result of dividing the population into two or three-dimensional grids. When the purpose of sampling is increasing the probability of locating possible hot spots in a population, systematic sampling is used.

In two-stage sampling, elementary units are randomly selected within the population and sample increments are taken from locations within each unit. The locations may be selected systematically or randomly.

Stratified random sampling is composed of the division of a population into sections called strata. By designing an efficient and cost-effective sampling plan, the number, size, and shape of strata are important. When the purpose of sampling is to estimate more precisely analyte concentration in the population, the uniformity of each stratum is necessary. So, the number of sample increments needed to define analyte distribution within each stratum will be reduced. By using systematic sampling, the units occur at the same relative position in the stratum, while in stratified random sampling, the position in the stratum is determined separately by randomization within each stratum.

In survey sampling, a group of distinct and identifiable units in the population is called a cluster and choosing a group of units as a single unit is called cluster sampling. In cluster sampling, the population is divided into small groups with each group serving as a sample unit. The formation of clusters means forming groups of a heterogeneous nature [6, 9].

5. Sampling errors

Sampling as the most crucial step of each analytical procedure is of high importance and should be done with the utmost care and expertise. Nevertheless, there are some typical errors (systematic and random) that can occur from a small percentage to several orders of magnitude [2]. The sampling error is the error caused by observing a sample instead of the whole population. The sampling error is the difference between a sample statistic used to estimate a population parameter and the actual but unknown value of the parameter. One of the most important systematic errors, which can occur in the sampling process, is sample contamination. Contamination can exceed the true levels of the compound analyzed and consequently causes to obtain invalid results. Therefore, one of the concerns of sampling is to avoid introducing contamination into the sample. The equipment for sample collection is one of the main sources of contamination in sampling so a proper sampling device and container are of great importance for accurate analysis. For example, in the sample storage step, by storing samples in containers a wide range of chemical or physical changes can occur such as chemical reactions within components of the sample or sample components reacting with the containers [6]. Another source of error in sampling is selecting species, which is representative of the sample. In many environmental samples, the concentration of analytes is relatively high, for example, at the $mg.kg^{-1}$ level and selecting species considered as bioindicators or biomonitors is decisive term. Also, some errors are caused by the homogenizing process to reduce initial values of species into manageable amounts in order to consider them as representative analytical subsamples. There is a relation between the sampling error and particle size of the sample as shown in **Figure 2**. As it can be seen, a larger sample size reduces error significantly; however, working with a large amount of sample is not logical. According to the figure, sampling error for a normal sample is likely to be between 0.1% and 1% of the whole sample. The sampling error can be minimized by grinding the particles into fine size with this assumption that there are no issues relating to the stability of the analytes. It is important to mention that the place of sampling is very closely linked to a proper sampling strategy. Carefully choosing of a statistically relevant number and mass of individual samples greatly depends on the distribution and concentration of the analyte in the collected material [2, 4].

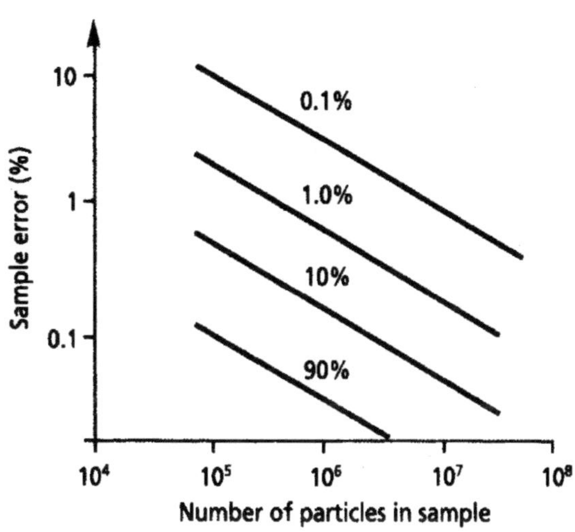

Figure 2.
The relationship between sample error, the number of particles in a sample, and different percentages of original material sampled [4].

6. Sample preparation techniques

Signal detection of the analytical instrument of the target analyte at a low concentration is always challenging because of the highly complicated media and different interferences from the matrix of any real sample such as biological, food, water, and environmental samples [10]. Accordingly, it is a need to develop a novel, effective, fast, inexpensive, and simple sample preparation technique to isolate analyte components of the samples and preconcentrate them to a detectable limit [11]. Several pretreatment techniques including solid phase extraction (SPE) [12], dispersive solid phase extraction (DSPE) [13], solid phase microextraction (SPME) [14], magnetic solid phase extraction (MSPE) [15], liquid–liquid extraction (LLE) [16, 17], molecular imprinted polymer (MIP) [18], stir bar sorptive extraction (SBSE) [19], pipette tip microsolid phase extraction [20], molecularly imprinted stir bar sorptive extraction (MIPSB) [21], microwave-assisted extraction (MASE) [22], membrane extraction [23], solid phase extraction (SPE) [24], supercritical fluid extraction (SFE) [25], and silver nanoparticle stir bar sorptive extraction [26] have been used widespread application for preconcentration and isolate trace level compound of complicated matrices. Before talking in detail about these mainly miniaturized techniques, first some details about the classification of extraction methods are explained.

7. Exhaustive and non-exhaustive extraction methods

Figure 3 represents the classification of extraction protocols. In exhaustive extraction methods such as solid phase extraction and liquid–liquid extraction, analytes are completely transferred to the extraction phase. These methods have been widely applied as a sample preparation technique in the separation and enrichment of inorganic and organic analytes of water media. Exhaustive extraction methods do not need calibration because almost all target molecules are transported into the extracting solvent because of their large volume. Flow systems replaced batch equilibrium techniques to reduce the consumption of solvent and time. For example, in dynamic extraction with solvent (Soxhlet extraction) in which extraction is performed at the solvent boiling point, circulation of the extractant enhances the extraction power significantly.

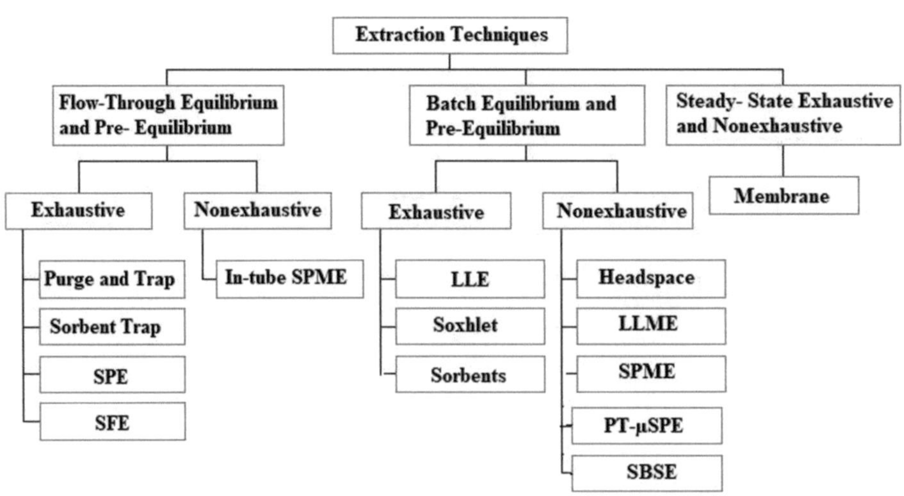

Figure 3.
Classification of extraction methods. Copyright Wiley-VCH, 2010. Reprinted with permission [1].

Non-exhaustive methods including SPME and liquid phase microextraction (LPME) are based on the principle of equilibrium, pre-equilibrium, and permeation. Although non-exhaustive methods (equilibrium method) such as SPME are similar to exhaustive equilibrium techniques, the amount of extraction phase is much smaller than equilibrium exhaustive techniques and is generally used to separate a small fraction of the analyte from the sample matrix [1].

8. Microwave and sonication extraction

Abu-Samra et al. [27] had already utilized the microwave method to promote the extraction techniques of liquid phase extraction (LPE). Microwave-assisted extraction can enormously decrease the extraction time (around 10 min) because microwaves directly heat up the solutions (sample and extraction solvent) instead of containers. The protocol also reduces the solvent consumption and allows multiple samples. The microwave-assisted extraction (MAE) effectiveness was proved with compared by general extraction techniques including high-speed homogenization and mechanical shaking protocols. MAE utilizes a household microwave oven that can reduce the cost of purchasing a dedicated tool for extraction, including a polytron-type extractor or an automated pressurized liquid extraction (PLE). If a flammable organic solvent including acetone is applied as an extraction solvent and a household microwave is utilized to MAE, sparks that may occur within the range can cause a fire or explosion. So, it is necessary to take suitable measures including as preventing the solvent from leaking into the extraction container to minimize fire and explosion danger [28, 29].

Ultrasonic-assisted extraction (UAE) or sonication extraction (SE) is a mature extraction methodology that accelerates the process of entering goal analytes in the solvent by utilizing ultrasound during liquid phase extraction (LPE). Indeed, ultrasonic extraction is based on mass exchange from the sample into the extraction phase using ultrasound radiation. Sonication extraction is done by high-frequency sound waves introduced into a liquid medium and it is associated with the formation and collapse of tiny bubbles or cavities in the liquid. It has been demonstrated that the UAE can improve the recovery compared to normal solvent extraction due to its physicochemical effects such as similar sonication cavitation, disturbance, fragmentation, emulsification, and erosion [30–32]. The extraction performance of UAE, MAE, and Soxhlet extraction was compared and the result showed the better performance of UAE [33]. The technique is a simple and cost-effective method [28].

9. Gas extraction techniques

Purge and trap protocol or dynamic headspace analysis or gas extraction technique is usually applied as an extraction method for the trace level determination of volatile analytes present in a liquid sample. The main advantage of the protocol is its capacity to accept large volumes of sample including the high sorption capacity of traps; hence, a lower detection limit for volatile organic analytes can be obtained. In the protocol, the sample is purged using inert gas for a specified time and the purged compounds are trapped applying an adsorbent trap as shown in **Figure 4**. Thermal desorption is utilizing to transfer the target molecules to a next step. Often before GC detection, an analyte accumulation step, depending on the cry-focusing unit, is applied. A short capillary is cooled to a sub-ambient temperature usually with liquid nitrogen or a Peltier system. The step allows to focus the compounds in a narrow band that is rapidly transferred to GC column using fast thermal desorption. This created a good peak shape and efficiency of separation, so it increases the resolution as well as the sensitivity of the procedure [34, 35].

10. Membrane extraction

Membrane extraction is an extraction technique that utilizes a microporous membrane to separate the analytes of the extractant. The materials in the matrix to be tested pass *via* the microporous membrane, which are then extracted using the extractant. Based on the physical and chemical properties of the matrix, the extractant can be adjusted and selected to achieve better detection results [36]. The concept of membrane extraction is shown in **Figure 5**.

Membrane extraction with a sorbent interface (MESI) is considered as an efficient extraction technique since it has not only the ability to isolate the compound of the real sample, but also indicates advantages such as high selectivity, removal of the extraction solvent, reduced analysis time, compatibility by gas chromatography (GC), and applicability to continuous monitoring and automation of industrial processes. So, the utilization of a MESI method for isolation and enrichment of target compounds is an area of growing research interest. The membrane extraction includes two simultaneous steps: the isolation of target compounds of the sample media using the membrane probe and the stripping of target molecules of the other side of the membrane utilizing a flowing stripping gas. The membrane acts as a selective barrier as it allows some compounds and prevents others from passing *via* the membrane wall to the analytical sample. MESI-GC has been used to isolate different samples in various media. A custom-made tools for heating the polymer and collection, isolate of the degradation products present in the sample's headspace was applied (**Figure 6**) [37]. The treated polymer film was sandwiched as a thin layer between two aluminum foils near the heating element, which was also covered by an aluminum foil. The temperature of the heating element was controlled with an automatic temperature controller. Also, the temperature of the heating element was determined precisely with a thermocouple wire, which was in contact with the aluminum foil and connected to the digital thermometer. A polymeric membrane

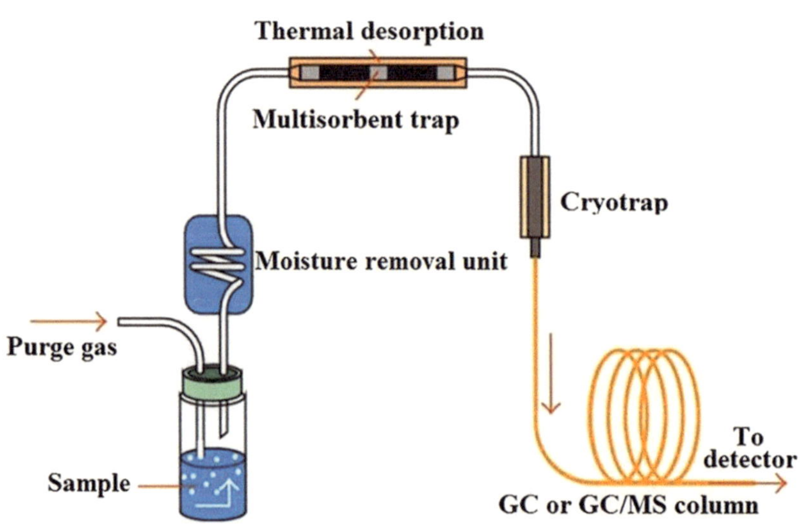

Figure 4.
Schematic diagram of purge and trap method (adapted from Ref. [34]). Reprinted with kind permission from Ref. [34] copyright 2020 Taylor & Francis [34].

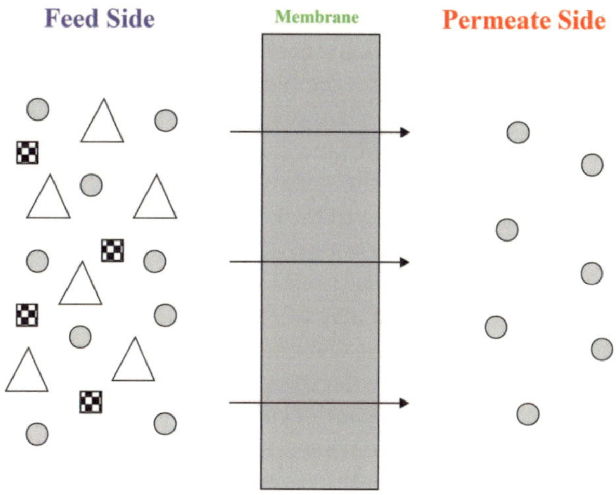

Figure 5.
The concept of membrane separation; the analytes are depicted by the circles [37]. Copyright Wiley-VCH, 2003. Reprinted with permission [37].

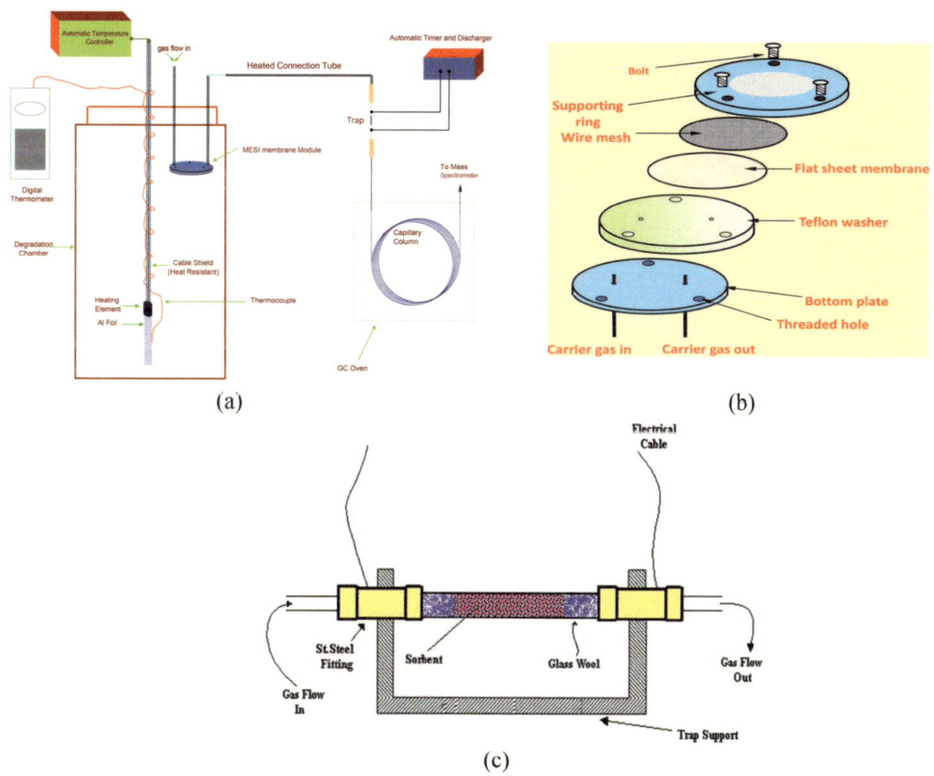

Figure 6.
Schematic diagram of MEAI-GC system applied for the polymer degradation and (a), details of the membrane module (b), main components of the MEAI trap (c). Reprinted by kind permission from Ref. [38]. Copyright 2002 Royal Society of Chemistry.

first extracts the compound from its media. By a porous adsorbent, the analytes are trapped on a polymeric trap. The extraction membrane acts as a selective barrier and is generally non-polar, then keeping water away from entering the system. It also acts as a selective element, since the permeation rates of various analytes vary by the membrane particle. Once the molecules cross the membrane, the carrier gas stream transports them to the adsorbent interface, where concentration occurs.

The MESI system was included of a membrane extraction probe, an adsorbent trap, and a capacitive discharge power supply. **Figure 6b** shows a scheme of the membrane module. A flat sheet membrane is not self-supportive and so needs a holder to support it for contact by the system. One side of the membrane was contacted to the sample and the other to the carrier gas. There are polytetrafluoroethylene (PTFE) washers and stainless steel plates on both sides of the membrane. In a stainless steel plate, one side of the membrane was supported with a mounted fine stainless steel wire mesh. Both plates were fastened using three bolts. *Via* small diameter PTFE tubing, the carrier gas was combined to the module, which fits tightly into the PTFE washer. The carrier gas pressure lifted the membrane and allowed free passage of the gas to the outlet tubing on the opposite side of the membrane surface. After leaving the membrane, the carrier gas and target molecules proceed to the small trap. The connecting tubes between the membrane module and the GC injector were deactivated stainless steel tubes. To delete the signal determined with the detector that could come of the compound absorbed on the walls of the tubes, the connectors were heated to approximately 120°C applying a heating tape. In **Figure 6c**, the design of the micro-trap was presented. The micro-trap was made by packing a section of stainless steel tubing by the polymer adsorbent. Using passing an electrical current directly *via* the wall of the tubing for a few milliseconds of the discharge, the trap was heated. The temperature was modified with an electrical potentiometer charged utilizing the capacitor and utilized to the ends of metal tubing. The cycle of trapping and heating was repeated automatically using a timer. The temperature of the trap was set producing minimal decomposition of the adsorbent [38].

11. Miniaturization in sample preparation and extraction

Miniaturization techniques are of increasing interest in almost all aspects of chemical analysis and especially in sample preparation. They lead to a significant decrease in analysis time, sample and solvent volume, and consumption of chemical reagents. Reduction in the quantity of chemical reagents and matrix samples applied per assay enables higher-throughput assays due to massive parallelization and multiplex determination versions and after that promulgates the creation of new procedures using easier handling protocols. Miniaturized SPE and LLE have been applied for enrichment and isolation of various analytes and have many advantages including low consumption of solvent, eluent, and sorbent, good extraction efficiency, simple operation versatility. These advantages led to the invention of miniaturized modes of SPE and LLE including SPME, SBSE, μSPE, hollow fiber-based liquid phase microextraction (HF-LPME), LPME, and DLLME [39–41].

Liquid-phase microextraction was first introduced by the Cantwell group and Dasgupta group in 1996 [42, 43]. As a sample preparation technique derived from LLE, LPME overcomes the drawbacks of solvent consuming and integrates sampling, enrichment, and extraction in one step. Compared with LLE, LPME has a much higher preconcentration factor and the method minimizes the solvent in scale of microliters. Various types of LPME have been employed to meet the requirements of determination such as HF-LPME, DLLME, and electromembrane extraction (EME). In HF-LPME, a porous hollow fiber by the immobilized solvent is placed between sample and acceptor solution for extraction, which provides high enrichment and extraction efficiency. The technique has advantages such as high sensitivity and sample cleanup [44].

DLLME has excellent enrichment characteristics and is fast. In DLLME, target compound extraction is practically instantaneous due to the enormous contact surface between the receptor and donating solvents. To obtain analyte dispersion, a third solvent should be applied. In DLLME, after adding a proper volume of extraction and dispersive solvent in the aquatic sample, a cloudy solution consisting of three solvents (water solution, extraction solvent, and dispersive solvent) is formed. After extraction of the target analyte from the water phase in fine droplets of extraction solvent, the organic phase can be separated by centrifugation [45, 46].

In EME, target molecules at aqueous media are electrically migrated across a solvent immobilized porous polymeric liquid membrane called supported liquid. They can later enter to the acceptor solution with the application of direct current electric potential. Furthermore, utilizing a transparent liquid free membrane as an alternative of SLM enables visual screening of the extraction methods. The EME is a promising procedure for wider applications in analyte analysis and efforts have been made to improve this extraction method [44].

Solid phase microextraction, as an effective and simple sample preparation, has been generally used for different real samples since it was proposed by Pawliszyn et al. in 1990 [47]. The extracting phase attached to the coated fiber has key effects on the extraction process of the method since the mechanism of the technique is mainly depending on the absorption equilibrium of the target compounds between the fiber coating particle and sample solution. Up to date, scientists have employed a number of coating particles to develop the usage of SPME including graphene oxide, covalent organic framework, polymer, molecularly imprinted polymer, ionic, and metal–organic frameworks (MOFs) [17].

Stir bar sorptive extraction is another interesting extraction method. Stir bar coated by poly-dimethylsiloxane (PDMS), for preconcentration and separation of the analytes, was first introduced in 1999 with Sandra et al. and called stir bar sorptive extraction (SBSE) [48]. The technique depends on sorption that makes target compounds partitioning between retaining sorbent (polymer) and liquid or vapor sample media; hence, originating bulk retention. SBSE is dipped in the sample and stirred for a defined time. After extraction of the target molecules, it is taken out of the sample and immerged in a vial including elution solvent and stirs again. In final, the elution solvent is removed and send to a proper apparatus for determination. Sampling is carried out with direct insertion of SBSE in liquid or headspace (HS) sample [49–51]. The mechanism of SBSE is shown in **Figure 7**.

Pipette tip microsolid phase extraction (PT-μSPE) is one of the most important extraction techniques of miniaturized SPE that has become a simple and suitable device for the extraction of different analytes. The pipette tip is proper to be applied

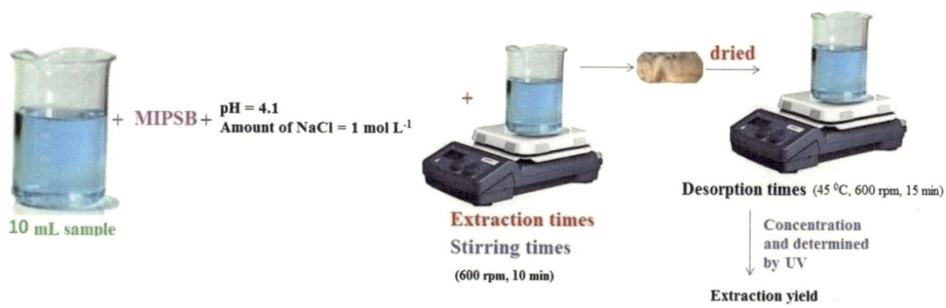

Figure 7.
Schematic of MIP-coated SBSE for naphthalene sulfonates. Reprinted by kind permission from Ref. [52]. Copyright 2018 Taylor and Francis.

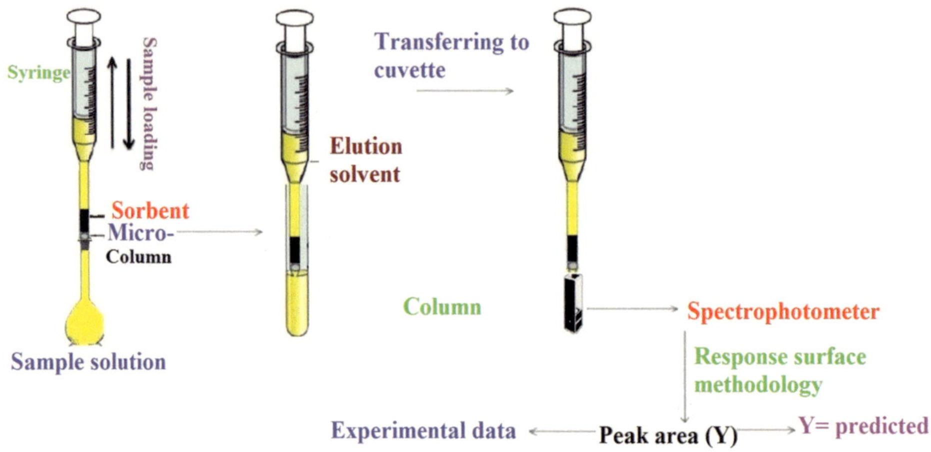

Figure 8.
Schematic of PT-μSPE for nalidixic acid and acetaminophen [40]. Reprinted by kind permission from Ref. [40]. Copyright 2020 springer.

Method	Extraction phase	Sample phase	Sampling mode
SBSE	Sorbent coated on stir bar	Environment or water sample	Direct
SDME	Single drop	Environment, water, and gas sample	Direct or headspace
DLLME	Organic solvent	Environment or water sample	Direct
LPME	Liquid phase	Environment or water sample	Direct
SPME	Polymeric extracting phase coated on a fused silica fiber	Environment, water, and gas sample	Direct, headspace, and membrane

Table 1.
Some of miniaturized sample preparation techniques.

as a SPE column due to its special conical shape using various diameters in two ends. Normally, a bigger pipette tip is immersed in a small one to obtain a novel cartridge and used as μSPE [53, 54]. Schematic of PT-μSPE for nalidixic acid and acetaminophen extraction is depicted in **Figure 8**.

Table 1 shows the comparison of some of the miniaturized sample preparation methods.

12. Automation

The aim of miniaturized methods is a drastic reduction in solvent consumption and samples, based on the green chemistry protocols. However, they sometimes consist of steps such as centrifugation, drawn and drop cycles, and vortexing which are regarded as disadvantages. Hence, even considering all non-automated micromethods benefits, they are still related to multiple steps, increasing the analysis time, complexity, and propensity of analytical errors. So, automated methods are explained basically to be simpler, faster, and further environmentally friendly than traditional and miniaturized methods. For the goal, various strategies for automation methods have been introduced in the last decades, such as the automation of SPE (online SPE) and with tailoring the microextraction protocols in order toward its automation such as automated SPME, in-tube SPME, automated SBSE, and so on. Fully automated SPME injection systems are commercially available now from different companies. It will be possible theoretically that a technique depending on SPME could be adapted to work by another microextraction technique, for example, by microextraction by packed sorbent (MEPS) once the same adsorbents such as C_{18} or C_8 can be packed inside the MEPS's syringe, which is applied as coated phase on SPME fibers. Hence, after preliminary tests to enough the values of the parameters including extraction cycles and solvent volume, the protocols could be transferred between these two methods and to other sorbent-based techniques as well [55].

13. Conclusion

Evaluating chemical analytes in complex media is a challenging task, which needs highly accurate sample preparation techniques. The development of these methods created miniaturization techniques with features such as providing lower detection limits, wider linear range, and improvement of selectivity and speed. Nevertheless, the intense sample handling remains an important gap that may even impair the performance of the techniques since the equilibrium between analytes can be easily disturbed. In this chapter, we tried to explain some major sample preparation techniques, both exhaustive and non-exhaustive methods. Also, sample collection and pretreatment are discussed as critical steps, which may impair the correct analytes detection despite the sample preparation method applied. Improvement of sample preparation using smaller samples, and the volume of solvent, reduction of analysis time, miniaturization, compatibility to different analytical instruments and automation, is still a major part of research in chemical analysis. Besides automation, the application of "green" solvents such as ionic liquids and deep eutectic solvents likely will find more demands in the extraction. Solventless and flow-through sample preparation methods are also of interest, which is expected to find more importance in the near future. Advances in membrane technology will affect extraction techniques as well. In the next chapters of this book, the reader will find some recent researches currently under investigation to achieve these goals in sample preparation.

Acknowledgements

We gratefully acknowledge team members for their essential contribution over the years.

Conflict of interest

The authors declare no conflict of interest.

Author details

Mona Sargazi[1], Sayyed Hossein Hashemi[2] and Massoud Kaykhaii[1*]

1 Faculty of Sciences, Department of Chemistry, University of Sistan and Baluchestan, Zahedan, Iran

2 Faculty of Marine Science, Department of Marine Chemistry, Chabahar Maritime University, Chabahar, Iran

*Address all correspondence to: kaykhaii@gmail.com

References

[1] Pawliszyn J, Lord HL. Handbook of Sample Preparation. 1st ed. New Jersey: Wiley; 2010. pp. 4-6. DOI: 10.1002/9780813823621

[2] Stoeppler M, editor. Sampling and Sample Preparation: Practical Guide for Analytical Chemists. 1st ed. New York: Springer-Verlag; 2008. p. 111 DOI: 001: 10.1007/978-3-642-60632-8

[3] Mitra S, Brukh R. Sample Preparation Techniques in Analytical Chemistry. 1st ed. New Jersey: Wiley; 2003. Chapter 1. pp. 1-30

[4] Ahuja S, Diehl D. Chapter 2 Sampling and sample preparation. Journal of Comprehensive Analytical Chemistry. 2006;**47**:15-40

[5] Hussain CM, Kecili R. Modern Environmental Analysis Techniques for Pollutants; Chapter 4, Sampling and sample preparation techniques for environmental analysis. 1st ed. Netherlands: Elsevier Inc.; 2020. pp. 75-119. DOI: 10.1016/B978-0-12-816934-6.00004-7.

[6] Pawliszyn J. Comprehensive Sampling and Sample Preparation: Analytical Techniques for Scientists. 1st ed. Netherlands: Elsevier; 2012. p. 3200

[7] Moldoveanu SC, David V. Sample Preparation in Chromatography. 1st ed. Netherlands: Elsevier Science; 2002. pp. 10-16.

[8] Ramos L. Chemical Analysis of Food: Techniques and Applications; Chapter 1- Basics and advances in sampling and sample preparation. 1st ed. Netherlands: Academic Press; 2012. pp. 3-24. DOI: 10.1016/B978-0-12-384862-8.00001-7

[9] Mester Z, Sturgeon R. Sample Preparation for Trace Element Analysis. 1st ed. Netherlands: Elsevier; 2003. p. 1338

[10] Li XS, Zhu GT, Luo YB, Yuan BF, Feng YQ. Synthesis and Applications of Functionalized Magnetic Materials in sSample Preparation. TrAC, Trends in Analytical Chemistry. 2013;**45**:233-247. DOI: 10.1016/j.trac.2012.10.015

[11] Jon CS, Meng LY, Li D. Recent review on carbon nanomaterials functionalized with ionic liquids in sample pretreatment application. TrAC, Trends in Analytical Chemistry. 2019;**120**:115641. DOI: 10.1016/j.trac.2019.115641

[12] Song Y, Ma R, Hao L, Yang X, Wang C, Wu Q, et al. Application of covalent organic framework as the adsorbent for solid-phase extraction of trace levels of pesticide residues prior to high-performance liquid chromatography-ultraviolet detection. Journal of Chromatography A. 2018;**1572**:20-26. DOI: 10.1016/j.chroma.2018.08.033

[13] Wang R, Li S, Chen DW, Zhao YF, Wu YN, Qi KM. Selective extraction and enhanced-sensitivity detection of fluoroquinolones in swine body fluids by liquid chromatography high resolution mass spectrometry: Application in long-term monitoring in livestock. Food Chemistry. 2021;**341**:128269. DOI: 10.1016/j.foodchem.2020.128269

[14] Celeiro M, Facorro R, Dagnac T, Llompart M. Simultaneous determination of trace levels of multiclass fungicides in natural waters by solid – phase microextraction-gas chromatography-tandem mass spectrometry. Analytica Chimica Acta. 2018;**1020**:51-61. DOI: 10.1016/j.aca.2018.03.014

[15] Huang C, Qiao X, Sun W, Chen H, Chen X, Zhang L, et al. Effective extraction of domoic acid from seafood based on postsynthetic-modified magnetic zeolite imidazolate framework-8 particles. Analytical

Chemistry. 2019;**91**:2418-2424. DOI: 10.1021/acs.analchem.8b05202

[16] Torbati M, Mohebbi A, Farajzadeh MA, Afshar Mogaddam MR. Simultaneous derivatization and air-assisted liquid-liquid microextraction based on solidification of lighter than water deep eutectic solvent followed by gas chromatography-mass spectrometry: An efficient and rapid method for trace analysis of aromatic amines in aqueous samples. Analytica Chimica Acta. 2018;**1032**:48-55. DOI: 10.1016/j.aca.2018.06.025

[17] Duo H, Lu X, Wang S, Liang X, Guo Y. Preparation and applications of metal-organic framework derived porous carbons as novel adsorbents in sample preparation. Trends in Analytical Chemistry. 2020;**133**:116093. DOI: 10.1016/j.trac.2020.116093

[18] Hashemi SH, Kaykhaii M, Khajeh M. Molecularly imprinted polymers for stir bar sorptive extraction: synthesis, characterization, and application. Analytical Letters. 2015;**48**:1815-1829. DOI: 10.1080/00032719.2014.1003431

[19] Hashemi SH, Kaykhaii M. Nanoparticle coatings for stir bar sorptive extraction, synthesis, characterization and application. Talanta. 2021;**221**:121568. DOI: 10.1016/j.talanta.2020.121568

[20] Hashemi SH, Kaykhaii M, Keikha AJ, Parkaz A. Application of box–behnken design in the optimization of a simple graphene oxide/zinc oxide nanocomposite-based pipette tip micro-solid phase extraction for the determination of rhodamine b and malachite green in seawater samples by spectrophotometry. Analytical Methods. 2018;**10**:5707-5714. DOI: 10.1039/C8AY02309C

[21] Hashemi SH, Kaykhaii M, Tabehzar F. Molecularly imprinted stir bar sorptive extraction coupled with high-performance liquid chromatography for trace analysis of naphthalene sulfonates in seawater. Journal of the Iranian Chemical Society. 2016;**13**:733-741. DOI: 10.1016/j.talanta.2011.03.041

[22] Kapoore RV, Butler TO, Pandhal J, Vaidyanathan S. Microwave-assisted extraction for microalgae: From biofuels to biorefinery. Biology. 2018;**7**:18-43. DOI: 10.3390/biology7010018

[23] Castro-Muñoz R, Díaz-Montes E, Cassano A, Gontarek E. Membrane separation processes for the extraction and purification of steviol glycosides: An overview. Critical Reviews in Food Science and Nutrition. 2020;**4**:1-23. DOI: 10.1080/10408398.2020.1772717

[24] Ötles S, Kartal C. Solid- phase extraction (SPE): Principles and applications in food samples. Acta Scientiarum Polonorum. Technologia Alimentaria. 2016;**15**:5-15. DOI: 10.17306/J.AFS.2016.1.1

[25] Ahmad T, Masoodi FA, Rather SA, Wani SM, Gull A. Supercritical fluid extraction: A review. Journal of Biological and Chemical Chronicles. 2019;**5**:114-122

[26] Hashemi SH, Kaykhaii M, Keikha AJ, Sajjadi Z, Mirmoghaddam M. Application of response surface methodology for silver nanoparticle stir bar sorptive extraction of heavy metals from drinking water samples: a box–behnken design. Analyst. 2019;**144**:3525-3532. DOI: 10.1039/C9AN00165D

[27] Abu-Samra A, Morris JS, Koirtyohann SR. Wet ashing of some biological samples in a microwave oven. Analytical Chemistry. 1975;**47**:475-1477. DOI: 10.1021/ac60358a013

[28] Chen X, Wu X, Luan T, Jiang R, Ouyang G. Sample preparation and instrumental methods for illicit drugs in environmental and biological samples: A review. Journal of Chromatography A. 1640;**2021**:461961. DOI: 10.1016/j.chroma.2021.461961

[29] Watanabe E. Review of sample preparation methods for chromatographic aanalysis of neonicotinoids in agricultural and environmental matrices: From classical to state-of-the-art methods. Journal of Chromatography A. 1643;**2021**:462042. DOI: 10.1016/j.chroma.2021.462042

[30] Lianfu Z, Zelong L. Optimization and comparison of ultrasound/microwave assisted extraction (umae) and ultrasonic assisted extraction (uae) of ly- copene from tomatoes. Ultrasonics Sonochemistry. 2008;**15**:731-737. DOI: 10.1016/j.ultsonch.20 07.12.0 01

[31] Zhang G, He L, Hu M. Optimized ultrasonic-assisted extraction of flavonoids from Prunella Vulgaris L. and evaluation of antioxidant activities in vitro. Innovative Food Science & Emerging Technologies. 2011;**12**:18-25. DOI: 10.1016/j.ifset.2010.12.003

[32] Mason TJ, Calinescu I. Ultrasonically assisted extraction (UAE) and microwave assisted extraction (MAE) of functional compounds from plant materials. TrAC Trends in Analytical Chemistry. 2017;**97**:159-178. DOI: 10.1016/j.trac. 2017.09.002

[33] Wilkinson JL, Hooda PS, Swinden J, Barker J, Barton S. Spatial (bio) Accumulation of pharmaceuticals, illicit drugs, plasticisers, perfluorinated compounds and metabolites in river sediment, aquatic plants and benthic organisms. Environmental Pollution. 2018;**234**:864-875. DOI: 10.1016/j.envpol.2017.11. 090

[34] Hashemi SH, Kaykhaii M, Mirmoghaddam M, Boczkaj G. Preconcentration and analytical methods for determination of methyl tert-butyl ether and other fuel oxygenates and their degradation products in environment: A review. Critical Reviews in Analytical Chemistry. 2020. DOI: 10.1080/10408347.2020.1753164

[35] Zwank L, Berg M, Schmidt TC, Haderlein SB. Compound-specific carbon isotope analysis of volatile organic compounds in the low-microgram per liter range. Analytical Chemistry. 2003;**75**:5575-5583. DOI: 10.1021/ac034230i

[36] Wang J, Huang LJ, Xu ZJ, Liu Y, Xu JJ, Chen HY. Rapid analysis of bacteremia by membrane extraction electrospray ionization mass spectrometry. Chinese Journal of Analytical Chemistry. 2020;**48**: 1315-1324. DOI: 10.1016/S1872-2040(20)60050-0

[37] Kou D, Mitra S. Sample Preparation Techniques in Analytical Chemistry. 1st ed. New Jersey: Wiley; 2003. Chapter 3. pp. 139-182

[38] Kaykhaii M, Sarafraz-Yazdi A, Chamsaz M, Pawliszyn J. Membrane extraction with sorbent interface-gas chromatography as an effective and fast means for continuous monitoring of thermal degradation products of polyacrylonitrile. Analyst. 2002;**127**:912-916. DOI: 10.1039/B202960J

[39] Hashemi SH, Kaykhaii M, Keikha AJ, Mirmoradzehi E, Sargazi G. Application of response surface methodology for optimization of metal– organic framework based pipette-tip solid phase extraction of organic dyes from seawater and their determination with HPLC. BMC

Chemistry. 2019;**13**:59-69. DOI: 10.1186/s13065-019-0572-0

[40] Hashemi SH, Kaykhaii M, Jamali Keikha A, Naruie N. Application of molecularly imprinted polymer pipette tip micro-solid phase extraction of nalidixic acid and acetaminophen from pills and seawater samples and their determination by spectrophotometry. Chemical Papers. 2020;**74**:4009-4023. DOI: 10.1007/s11696-020-01215-0

[41] Hashemi SH, Keykha F. Application of the response surface methodology in the optimization of modified molecularly imprinted polymer based pipette-tip micro-solid phase extraction for spectrophotometric determination of nicotine in seawater and human plasma. Analytical Methods. 2019;**11**:5405-5412. DOI: 10.1039/C9AY01496A

[42] Jeannot MA, Cantwell FF. Solvent microextraction into a single drop. Journal of Analytical Chemistry. 1996;**68**:2236-2240. DOI: 10.1021/ac960042z

[43] Liu H, Dasgupta PK. Analytical chemistry in a drop. Solvent extraction in a microdrop. Journal of Analytical Chemistry. 1996;**68**:1817-1821. DOI: 10.1021/ac960145h

[44] Lia N, Zhang T, Chen G, Xu J, Ouyang G, Zhu F. Recent advances in sample preparation techniques for quantitative detection of pharmaceuticals in biological samples. TrAC Trends in Analytical Chemistry. 2021:116318. DOI: 10.1016/j.trac.2021.116318

[45] Nassiri M, Kaykhaii M, Hashemi SH, Sepahi M. Spectrophotometric determination of formaldehyde in seawater samples after in-situ derivatization and dispersive liquid-liquid microextraction. Iranian Journal of Chemistry and Chemical Engineering. 2018;**37**:89-98

[46] Rezaee M, Assadi Y, Hosseini MM, Aghaee E, Ahmadi F, Berijani S. Determination of organic compounds in water using dispersive liquid-liquid microextraction. Journal of Chromatography A. 2006;**1116**:1-9. DOI: 10.1016/j.chroma.2006.03.007

[47] Arthur CL, Pawliszyn J. Solid phase microextraction with thermal desorption using fused silica optical fibers. Journal of Analytical Chemistry. 1990;**62**:2145-2148

[48] Baltussen E, Sandra P, David F, Cramers C. Stir bar sorptive extraction (SBSE), a novel extraction technique for aqueous samples: Theory and principles. Journal of Microcolumn Separations. 1999;**11**:737-747

[49] Hashemi SH, Monfaredzadeh Z. Molecularly imprinted stir bar sorptive extraction coupled with high-performance liquid chromatography for trace analysis of diclofenac in different real samples. Iranian Journal of Chemistry and Chemical Engineering. 2019;**38**:173-183

[50] Hashemi SH, Najari F. Response surface methodology of pre-concentration of chorophenols from seawater samples by molecularly imprinted stir bar sorptive extraction combined with hplc: Box–behnken design. Journal of Chromatographic Science. 2019;**57**:279-289

[51] Hashemi H, Khajeh M, Kaykhaii M. Molecularly imprinted stir bar sorptive extraction coupled with atomic absorption spectrometry for trace analysis of copper in drinking water samples. Analytical Methods. 2013;**5**:2778-2783

[52] Hashemi SH, Kaykhaii M, Tabehzar F. Spectrophotometric determination of four naphthalene sulfonates in seawater after their molecularly imprinted stir bar sorptive

extraction. Journal of the Chilean Chemical Society. 2018;**63**:4032-4038

[53] Hashemi SH, Kaykhaii M, Keikha AJ, Mirmoradzehi E. Box-behnken design optimization of pipette tip solid phase extraction for methyl orange and acid red determination by spectrophotometry in seawater samples using graphite based magnetic $NiFe_2O_4$ decorated exfoliated as sorbent. 2019;**213**:218-227. DOI: 10.1016/j.saa.2019.01.049

[54] Hashemi SH, Kaykhaii M, Keikha AJ, Mirmoradzehi E. Box-behnken design optimization of pipette tip solid phase extraction for methyl orange and acid red determination by spectrophotometry in seawater samples using graphite based magnetic NiFe2O4 decorated exfoliated as sorbent. Spectrochimica Acta Part A: Molecular and Biomolecular Spectroscopy. 2019;**213**:218-227. DOI: 10.1016/j.saa.2019.01.049

[55] Maciel EVS, Medina DAV, Borsatto JVB, Lanças FM. Towards a universal automated and miniaturized sample preparation approach. Sustainable Chemistry and Pharmacy. 2021;**21**:100427. DOI: 10.1016/j.scp.2021.100427

Section 2

Sample Preparation Techniques for Surface Analysis

Chapter 3

Advanced Sample Preparation Techniques for Surface Spectroscopy Analysis of Organic: Inorganic Hybrid Silica Particles

Harekrishna Panigrahi, Smrutirekha Mishra and Suraj Kumar Tripathy

Abstract

Silica due to its large inorganic amorphous wall and hydrophilic surface properties renders its suitability for designing different varieties of organic–inorganic silica-based materials. Characterization of such hybrid silica-based materials is one of the fascinating as well as challenging topics to be covered. Surface analysis of these hybrid materials can be done utilizing various techniques, out of which X-ray photoelectron spectroscopy (XPS), ^{29}Si Solid-state Nuclear magnetic resonance (NMR) spectroscopy, and Fourier-transform infrared spectroscopy (FTIR) is the most ideal ones. Thus, before analyzing these silica materials, it requires a massive study on its sample preparation for appropriate characterization of the organic molecules present in the inorganic network. Hence, this chapter will give a brief elucidation of the sample preparation techniques for analyzing the hybrid materials utilizing the above instrumentation techniques.

Keywords: Silica, Surface analysis, Sample preparation, XPS, ^{29}Si Solid-state NMR, FTIR

1. Introduction

Characterization of solid materials [1] is a vast and intense field that requires immense knowledge and practice to obtain the required properties. Other than characterization two important challenges that any solid sample require are understanding the results obtained and second is sample or specimen preparation techniques [2]. In the last few decades, exhaustive research has been carried out in understanding these solid sample's composition, chemistry [3], and its internal structure [4] to push these samples towards an efficient approach for specimen preparation and analysis.

Out of all the solid samples such as in powdered form [5] or film form [6], studying these hybrid silica materials has been a keen focus for many researchers. Silica materials [7] due to their rigid inorganic structure [8] make the material an overall highly amorphous [9] bulk powdered form. Thus, focusing on these organic–inorganic hybrid silica [10] particles requires more attention due to the presence of organic moieties [11] in the inorganic network. As the chances of exchanges of ions/molecules [12] and redistribution of molecules [13] are higher when it undergoes different techniques of specimen preparation for its analysis as compared to other materials. The Exchange of ions/molecules is higher during the sample preparation technique due to the presence of impurities within the sample chamber. Sometimes it can lead to adsorption [14], vaporization [15], and corrosion on these powered surfaces. Due to its robust nature, the important parameter which needs to be king-pinned during its sample preparation is that the hybrid surface is majorly exposed to the probe of the characterization technique [16] rather than the contaminated one. Hence, in simple words maintaining the surface and cleaning it with very basic revision is the important challenge for its sample preparation technique. Thus, it is very important initially to understand the scope of the sample/specimen preparation technique prior to analyzing.

In this chapter we are focusing on the aspects of advanced sample preparation techniques that are required for analyzing these hybrid materials utilizing XPS [17], ^{29}Si Solid-state NMR [18], and FTIR [19] techniques. The very basic steps that are required for sample preparation utilizing all of these techniques are described in **Figure 1**.

One of the important features for these materials that are required for analysis utilizing surface characterization is the use of a very high vacuum system [20] (ultrahigh vacuum, 10^{-7}–10^{-11}) since these materials involve the detection of charged particles [21]. Hence, the analyst should be aware that whether the surface of the sample is susceptible to alternation or degradation in the vacuum system. In some cases, differentially pumped sample holders can be used which can keep the surfaces stable even at high vacuum pressure. Owing to the sample preparation technique one must always remember that what vacuum should be used for processing the hybrid materials. Some of the very important information which should be focused on while analyzing these hybrid materials utilizing surface characterization techniques are shown described in **Figure 2**.

Hence, this chapter is organized in different sections which will give an overall idea of the traditional and advanced sample preparation techniques which are utilized for surface spectroscopy analysis (XPS, ^{29}Si Solid-state NMR, FTIR) for organic–inorganic hybrid silica samples.

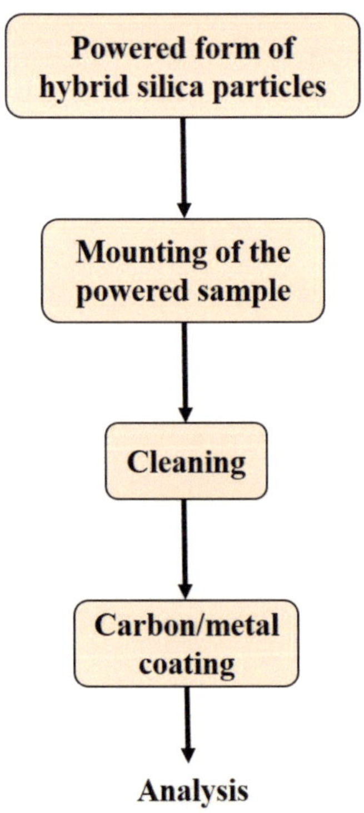

Figure 1.
Basic steps required for specimen preparation-XPS, ^{29}Si solid-state NMR, and FTIR.

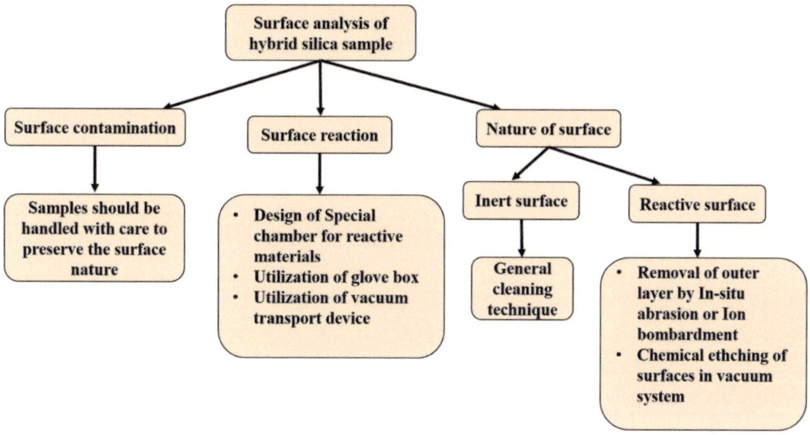

Figure 2.
Basic handling steps required for surface analysis of hybrid silica samples.

2. Surface analysis techniques

Characterizing these organic–inorganic silica particles [22] is an interesting as well as challenging topic. Though these materials contain rigorous inorganic silica [23] network within them still the challenge remains to identify these organic molecules within this rigid inorganic network. Hence, characterizing and analyzing these sample surfaces requires different techniques. Some of the important techniques which are largely used are described in **Table 1**.

Techniques	Concerns
Auger electron spectroscopy (AES)	High-resolution analysis of surface and film from ~1 to 20 Å depth.
X-ray photoelectron spectroscopy (XPS)	Depth profiling of surfaces and films up to 1-3 μm. Determines individual chemical composition and chemical state of elements.
Ion scattering spectroscopy (ISS)	Provides general elemental information and determines usually the monolayer structure of surfaces and films.
Secondary-ion mass spectroscopy (SIMS)	An ultra-high sensitivity technique that provides a piece of qualitative information on the elemental composition of surfaces and films
Wavelength dispersive spectroscopy (WDS)	Provides both qualitative and quantitative information on the elemental composition of surfaces and films with a detection limit of ~0.2%.
Energy dispersive spectroscopy (EDS)	Provides qualitative and quantitative along with elemental mapping with a detection limit of ~0.1%.

Table 1.
Surface analysis techniques are largely used for the analysis of organic–inorganic silica samples.

3. Traditional sample preparation techniques

Every sample preparation technique that is used for analyzing the surfaces of hybrid silica materials has its advantages [24] and disadvantages [25]. As the analysis techniques become more sophisticated, sample preparation becomes more and more precise [26]. However, focusing on surface analysis [27] of these hybrid samples they have special concerns. Since these hybrid materials have the presence of organic moieties on the outer surface which in return makes it more sensitive and fragile for sample preparation before analyzing through any technique.

Challenges which is concerned for these types of materials are (a) exchange of ions and atoms in the sample preparation environment and (b) redistribution and reconstruction of atoms/molecules within this bulk powdered sample. Other than these, one of the important parameters for analyzing hybrid surfaces is that they require an ultrahigh vacuum environment ($10^{-7} – 10^{-11}$ torr) as they require detection of charged silica particles.

3.1 X-ray photoelectron spectroscopy (XPS)

XPS [28, 29] is one of the sensitive characterization techniques which requires extreme alertness while its sample preparation [30]. One of the vital points which need to be focused on first is its sample handling practice. The important highlights for handling the samples are as follows:

- Powder-free nitrile or polyethylene gloves should always be used. Tweezer or glassware if used should thoroughly be cleaned before use.

- Cleaning of the sample chamber preferably by Ar^+ etching should be done to remove oxide layers that will cause damage to the surface of the silica samples.

- Since these hybrid materials have a high surface area, they require a longer time for pumping before positioning for analysis.

- These materials are highly sensitive to moisture/oxygen thus these should be loaded and mounted onto the vacuum transfer vessel in the glove box.

Generally, for XPS analysis the sample thickness should not be more than 4 mm. Since these hybrid materials are powder in form thus the general methods that are followed for its sample preparation along with its drawbacks are as follows:

- **Pressing of the powdered silica form into clean indium foil (high purity, as shown in Figure 3).** One problem which arises with this method; since Indium is a high atomic number (Z) material thus chances of spectral contamination with many XPS peaks are possible.

- **Utilization of a mesh screen for pressing finely powdered silica samples (as shown in Figure 4).** The Only drawback with this method is the chances of the destruction of organic groups present on the outer surface due to high mechanical work. There is also the possibility of the irradiating beams hitting the mesh which can lead to the destruction of some of the groups. Sometimes if the powered silica sample is not properly crushed entrapment of coarse particles in the mesh is possible which would ultimately not allow the passage of finer silica particles.

- **Preparation of pellets from powdered silica samples (as shown in Figure 5).** This is one of the advantageous methods for the preparation of samples. However, chances of potential structural changes due to the pressure applied are possible.

- **Placing of hybrid silica sample within the depression of the sample holder.** The major drawback of this technique is the potential loss of samples when high vacuum pressure is applied.

Figure 3.
Demonstration of the method used for pressing powdered silica samples onto indium foil [31].

Figure 4.
Demonstration of the method used for pressing powdered silica samples onto mesh screens [32].

Figure 5.
Demonstration of the method used for the preparation of pellets from powdered silica samples [33].

- **Placing of hybrid silica sample in the container and then tilting of the container towards the sample analyzer.** High chances of spilling silica samples within the vacuum chamber are possible when the alignment of the sample holder is performed.

3.2 ^{29}Si solid-state NMR

^{29}Si Solid-state NMR [34, 35] is a special NMR spectroscopy technique specifically used for the study of silica/silicon samples [36]. This technique can provide a depth of knowledge and understanding about the organic–inorganic hybrid silica samples [37]. However, sample preparation for this technique is purely different than standard solution NMR. Some of the important features of ^{29}Si Solid-state NMR are as follows:

Probes: Different types of probes [38, 39] are used for ^{29}Si Solid-state NMR which are efficiently capable of handling high power running systems (as shown in **Figure 6**).

Experiment: Parameters for ^{29}Si Solid-state NMR are quite different than the solution NMR. Such as chemical shift anisotropy (CSA) and dipolar coupling are usually averaged out in solution NMR due to the random motion. However, in the case of ^{29}Si Solid-state NMR CSA and dipolar coupling are quite dominating providing an entirely different experimental setup. Thus, one should have a piece of deep theoretical knowledge on the user's part.

Generally, sample preparation in the case of ^{29}Si Solid-state NMR is performed using specially designed "rotors" of different diameters usually made of zirconia. These rotors and their setup are highly expensive. These rotors packed with silica samples are spun at high frequency and a particular angle. This spinning of rotors at a magic angle (ca. 54.74°, where $\cos 2\theta m = 1/3$) concerning the magnetic field applied is referred to as magic angle spinning (MAS). These rotors are tube-like structures [40] used for MAS are quite small often white. There are different types of rotors of different capacity used for ^{29}Si Solid-state NMR such as (**Figure 7**):

Figure 6.
A general model of ^{29}Si solid-state NMR probe [39].

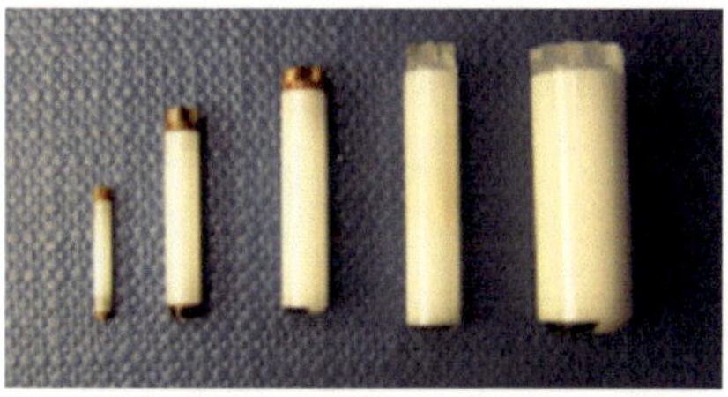

Figure 7.
Rotors of different capacities are used for sample packing in ^{29}Si solid-state NMR [40].

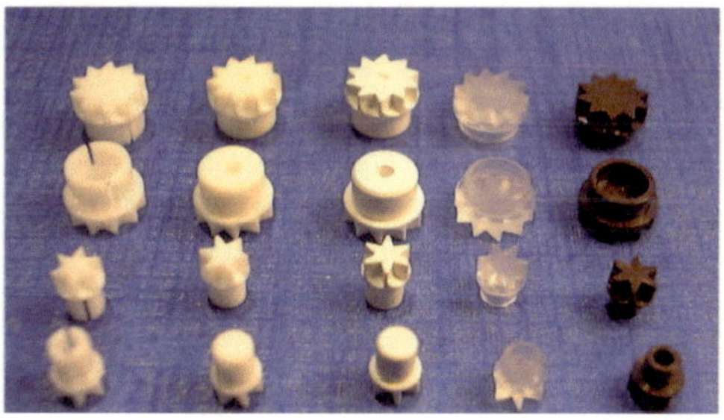

Figure 8.
Type of caps used for rotors (typical materials used ZrO_2, Macor, Kel and Vespel) [40].

These rotors are closed with caps after the hybrid silica samples are packed in them with caps. These caps are also of different sizes according to the rotor system and are made of different materials. Commonly used materials are Vespel, Ke1-F, Zirconia, and Boron nitride (**Figure 8**).

These hybrid silica samples are usually agglomerated powered in form. Thus, it is always recommended to grind these samples to a fine powder using mortar and pestle before inserting them into the rotor system. Then keep the rotors in a standing position and pack the silica samples in stages using a clean spatula. Filling of sample is required up to the cap's place.

While disassembling and removing the powered silica samples from the rotor one must be extremely careful in the opening of the system without breaking them apart as these are highly expensive. For example, in the case of the Ke1-F cap cooling of the rotor is first required to reduce the risk of breaking the cap. In the case of hybrid silica, samples remove samples from the rotor using a spatula. Then all of the parts should be washed properly preferably by a solvent (acetone) and dried in an oven.

3.3 Fourier-transform infrared spectroscopy (FTIR)

FTIR [41, 42] is a simple process that operates through transmission technique [43] and it does not require a separate accessory for analysis. It allows the users to simply place the powdered hybrid silica samples onto the infrared (IR) [44] beam setup. Since the IR beam passes through the silica samples, the transmitted energy is recorded and a final spectrum is obtained. However, there are different techniques for preparing samples specifically for silica materials. One very old method of preparing silica samples for analysis is *via* pressed pellet/KBr pellets and disks. To prepare pellets using silica sample and KBr, one must follow some rules such as:

- **Sample/KBr ratio:** A proper concentration of silica samples is required to be mixed with KBr powder to get a clear spectrum. Usually, between 0.2**–1%** of silica samples should be used to obtain a clear pellet.

Figure 9.
Pistil setup along with stainless steel disks for sample preparation of hybrid silica materials for analysis utilizing FTIR technique [45].

- **Sample preparation:** A homogenous mixture between the silica sample and KBr powder is expected to obtain the best results. The hybrid silica sample and KBr powder absorb humidity very fast (due to hygroscopic nature) from the environment which would eventually lead to increased background noises in the spectrum. Thus, drying of silica samples as well as KBr (preferably at 100°C) is highly required. After drying the materials take nearly 1 to 2% of the sample along with KBr powder and mix it properly using mortar and pestle [45] as shown in **Figure 9**. As these hybrid silica samples are a little bit hard thus first silica samples should be added and ground followed by the addition of KBr.

  Clutch two stainless steel disks and transfer the silica samples carefully onto one disk with a spatula that is inside the pistil. Then put another disk on top of the sample. Finally, transfer the pistil setup to the hydraulic press. Put a maximum pressure of 20,000 prf. Wait for few seconds so that the powered silica samples bind properly with the KBr and form a uniform pellet. Release the pressure and carefully remove the pistil out of the hydraulic press system. Slowly open the pistil and take out the pellet formed. The pellet should look homogenous and clear (mostly transparent) in nature.
  Sometimes, very uneven cloudy pellets are formed due to some of the reasons such as improper mixing of KBr powder, moisture in silica sample, sample to KBr ratio is quite high and pellet too thick.

4. Modern sample Preparation techniques

Due to increased demand for purity of samples leading to acceptable results are currently the hot topics; thus an advanced form of sample preparation technique is highly required. Sample preparation has always been a crucial factor for surface analysis techniques leading to most labor-intensive parts. This in turn makes an analysis slower and increases its cost leading to multi-step procedures. Usually, an ideal sample preparation technique should be simple, efficient, and inexpensive. The most important factors that have increased the curiosity in researchers for modern sample preparation techniques are increasing the sample loading capacity, decrease in labor cost, and good data collection. Hence, this section will provide an overview of the advanced sample preparation techniques are currently used for surface analysis using XPS, ^{29}Si Solid-state NMR, and FTIR specifically for organic–inorganic hybrid silica samples.

4.1 X-ray photoelectron spectroscopy (XPS)

As discussed above some of the traditional methods have been used for ages for preparing samples. These traditional methods lacked proper utilization of samples sometimes leading to loss of samples and time consumption. Highlighting these points researchers tried to design an advanced XPS system where sample preparation along with analysis would be much more efficient. Some of the advanced sample preparation techniques for XPS is as follows:

- Recently, a multi-function XPS having load-locked advanced sample preparation stages was developed which has additional ultraviolet photoelectron spectroscopy along with sample pre-treatments units. This XPS system has an in-situ sample preparation stage. This system has additional chambers attached along with the analysis vessel allowing loading of more than one sample at a single time. As this technique is quite advanced thus predictions of organic groups in the inorganic silica samples would be possible to measure quantitatively. Also, one must not follow the long procedures of the traditional method of sample preparation technique saving time as well as cost.

- One more advanced technique which has attracted quite an attention recently is a custom-made multi-chamber ultra-high vacuum [46] (UHV) system as shown in **Figure 10** for easy sample transport along with additional characterization ability. Its auxiliary system allows these silica samples to transport to the analysis vessel at very low pressure ($10-^{10}$ mbar). It also has a central radial distribution channel where more than one sample can be kept in the second sample preparation chamber or sample storage chamber when one is being analyzed. The sample can be heated and cooled during its measurements and these chambers are equipped with gas dosing for in-situ sample preparation. It has a total of six sample holders in the storage chamber.

  The preparation chambers provide in-situ sample preparation facilities such as sputtering and annealing which allows in-situ characterization of samples. These chambers are equipped with the precision manipulator, electron beam evaporator, pumping system, pressure gauges, and ions sources for surface cleaning. Since these have numerous UHV systems which allow the analysis of powdered silica samples having different surfaces.

4.2 ^{29}Si solid-state NMR

Hybrid silica samples require special analytical methods that can allow these samples to be analyzed both on macroscopic as well as atomic levels to co-relate the structure and property practically. Thus, ^{29}Si Solid-state NMR is a highly efficient technique for providing a wealth of information about these materials. As described above sample preparation techniques that are used traditionally have their owns advantages and disadvantages. However, an upgraded version for the sample preparation technique is always required to efficiently use these silica samples. As

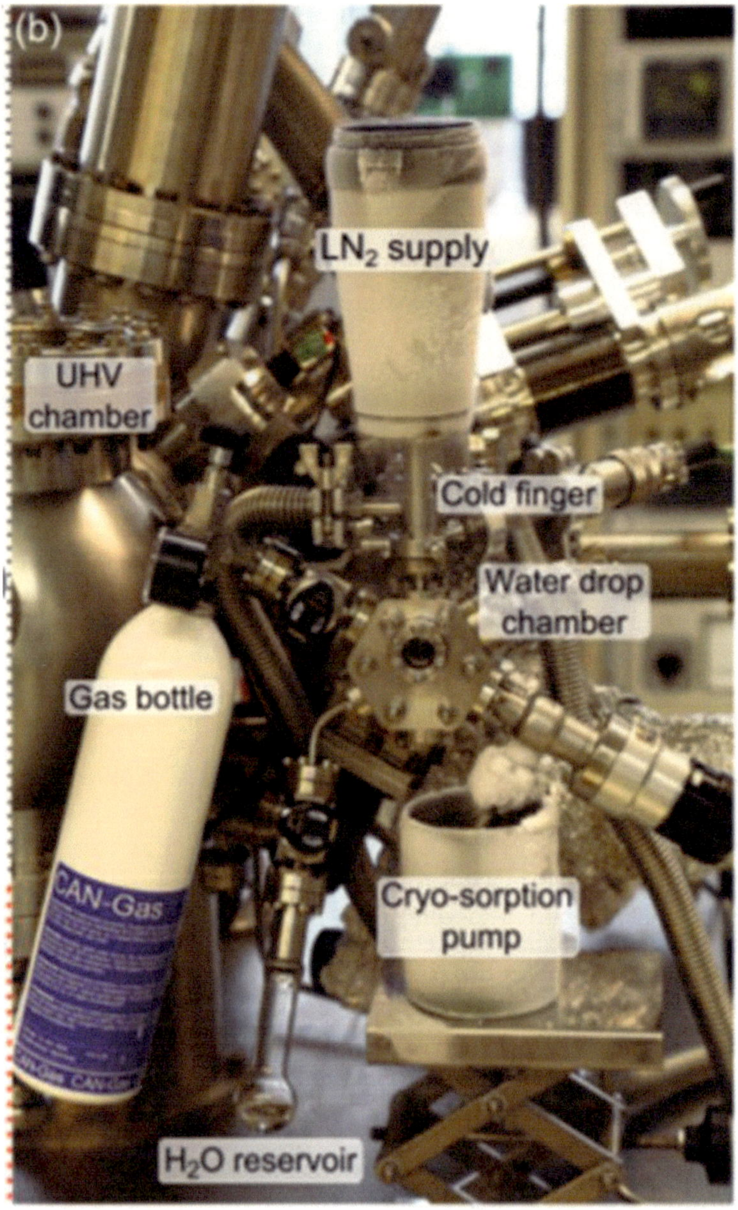

Figure 10.
A typical in-situ sample preparation chamber is attached to the UHV system XPS [46].

described above rotors play an important role in sample preparation and packing in ^{29}Si Solid-state NMR. Thus the efficiency of these rotors while spinning depends on the type of materials it is made up of. Hence, new types of rotors are now being introduced into the market with improved efficiency. Kel-F rotors are now upgraded to Torlon based rotors. Torlon is a green thermoplastic poly-imide-based material that has very strong chemical resistance (**Figure 11**).

As these hybrid silica samples are quite moisture/air sensitive materials thus it requires set-up which would allow the least moisture contact. O-ring caps are the best option for such types of materials. Macor rotors are also more preferred for silica-based samples. Usually, these O-ring caps are best suited for macor rotors.

Figure 11.
A general torlon based rotor was used for $^{\text{the 29}}$Si solid-state NMR technique [38].

Figure 12.
Typical setup of FTIR for sample preparation for both traditional and modern methods [47].

4.3 Fourier-transform infrared spectroscopy (FTIR)

The most common method for sample preparation in FTIR for hybrid silica samples involves grinding of sample into fine powdered form and then dispersing it into a matrix. KBr is the most widely used matrix material. As discussed above, initially drying of silica samples as well as KBr (preferably at 100°C) is highly required. After drying the materials addition 1 to 2% of the sample along with KBr powder and mix it properly using mortar and pestle. As these hybrid silica samples are a little bit hard thus first silica samples should be added and ground followed by the addition of KBr. Then clutch two stainless steel disks and transfer the silica samples carefully onto one disk with a spatula that is inside the pistil. Then put another disk on top of the sample. Finally, transfer the pistil setup to the hydraulic press. Put a maximum pressure of 20,000 prf. Wait for few seconds so that the powered silica samples bind properly with the KBr and form a uniform pellet. Release the pressure and carefully remove the pistil out of the hydraulic press system. Slowly open the pistil and take out the pellet formed. The pellet should look homogenous and clear (mostly transparent) in nature.

This is a universal method used always for sample preparation techniques in FTIR as this method is quite advantageous. Hence, an advanced sample preparation technique is not required for this technique (**Figure 12**).

5. Conclusions

Observing all the procedures for sample preparation utilizing different techniques, a detailed explanation is provided about the hybrid silica samples and their handling procedure before its sample preparation. As these materials have large inorganic amorphous walls and hydrophilic surface properties thus characterization of such hybrid silica-based materials is one of the fascinating as well as challenging topics to be covered. Surface analysis of these hybrid materials can be done utilizing various techniques, out of which X-ray photoelectron spectroscopy (XPS), ^{29}Si Solid-state Nuclear magnetic resonance (NMR) spectroscopy and Fourier-transform infrared spectroscopy (FTIR) are the most ideal ones. Thus, before analyzing these silica materials, it requires a massive study on its sample preparation for appropriate characterization of the organic molecules present in the inorganic network. Hence, this chapter provided a brief elucidation on the sample preparation techniques for analyzing the hybrid materials utilizing the above instrumentation techniques utilizing both traditional and advanced methods. Nevertheless, many approaches for sample preparation have been developed recently that would allow easy investigations of the target surfaces. In order to highlight the surface analysis of organic–inorganic hybrid materials implementation of new methods such as chip-based systems, optical tweezers, and micro-fluid-based systems would provide high potentiality to this technique. The extensive study still needs to be performed specifically for surfaces analysis that would push the traditional sample preparation methods towards high outturn measurements.

Author details

Harekrishna Panigrahi[1], Smrutirekha Mishra[2*] and Suraj Kumar Tripathy[2]

1 Department of Metallurgical and Materials Engineering, IIT, Patna, Bihar, India

2 School of Chemical Technology, KIIT University, Bhubaneswar, Odisha, India

*Address all correspondence to: smrutirekha.mishra@kiibiotech.ac.in

References

[1] E.A. Shneour, C.E. Ballou, The Preparation and Characterization of Di-O-benzylidene-(-)-inositol, J. Am. Chem. Soc. 80 (1958) 3960-3961. doi:10.1021/ja01548a040.

[2] L. Xia, J. Yang, R. Su, W. Zhou, Y. Zhang, Y. Zhong, S. Huang, Y. Chen, G. Li, Recent Progress in Fast Sample Preparation Techniques, Anal. Chem. 92 (2020) 34-48. doi:10.1021/acs.analchem.9b04735.

[3] J. Yan, B.K. Teo, N. Zheng, Surface Chemistry of Atomically Precise Coinage-Metal Nanoclusters: From Structural Control to Surface Reactivity and Catalysis, Acc. Chem. Res. 51 (2018) 3084-3093. doi:10.1021/acs.accounts.8b00371.

[4] J. Jeevanandam, A. Barhoum, Y.S. Chan, A. Dufresne, M.K. Danquah, Review on nanoparticles and nanostructured materials: History, sources, toxicity and regulations, Beilstein J. Nanotechnol. 9 (2018) 1050-1074. doi:10.3762/bjnano.9.98.

[5] C.F. Holder, R.E. Schaak, Tutorial on Powder X-ray Diffraction for Characterizing Nanoscale Materials, ACS Nano. 13 (2019) 7359-7365. doi:10.1021/acsnano.9b05157.

[6] X.J. Cai, H.M. Yuan, A. Blencowe, G.G. Qiao, J. Genzer, R.J. Spontak, Film-Stabilizing Attributes of Polymeric Core-Shell Nanoparticles, ACS Nano. 9 (2015) 7940-7949. doi:10.1021/acsnano.5b00237.

[7] F. Catalano, P.P. Pompa, Systematic Study, (2019). doi:10.1021/acsami.9b16135.

[8] K. Möller, T. Bein, Talented mesoporous silica nanoparticles, Chem. Mater. 29 (2017) 371-388. doi:10.1021/acs.chemmater.6b03629.

[9] A. Morikawa, H. Yamaguchi, M. aki Kakimoto, Y. Imai, Formation of Interconnected Globular Structure of Silica Phase in Polyimide-Silica Hybrid Films Prepared by the Sol-Gel Process, Chem. Mater. 6 (1994) 913-917. doi:10.1021/cm00043a009.

[10] G.B. Alexander, W.M. Heston, R.K. Iler, The solubility of amorphous silica in water, J. Phys. Chem. 58 (1954) 453-455. doi:10.1021/j150516a002.

[11] J. Wen, G.L. Wilkes, Organic/inorganic hybrid network materials by the sol-gel approach, Chem. Mater. 8 (1996) 1667-1681. doi:10.1021/cm9601143.

[12] A. Clearfield, Role of Ion Exchange in Solid-State Chemistry, Chem. Rev. 88 (1988) 125-148. doi:10.1021/cr00083a007.

[13] Z. Yang, M. Corso, R. Robles, C. Lotze, R. Fitzner, E. Mena-Osteritz, P. Bäuerle, K.J. Franke, J.I. Pascual, Orbital redistribution in molecular nanostructures mediated by metal-organic bonds, ACS Nano. 8 (2014) 10715-10722. doi:10.1021/nn504431e.

[14] H. Swenson, N.P. Stadie, Langmuir's Theory of Adsorption: A Centennial Review, Langmuir. 35 (2019) 5409-5426. doi:10.1021/acs.langmuir.9b00154.

[15] G. Marston, 47 1 47 1 •, 2267 (1991) 6-7.

[16] M.D. Montano, B.J. Majestic, A. Jamting, P.K. Westerhoff, J.F. Ranville, Methods for the detection and characterization of silica colloids by microsecond spICP-MS Methods for the detection and characterization of silica colloids by microsecond spICP-MS, (2016). doi:10.1021/acs.analchem.5b04924.

[17] M.T. Camci, B. Ulgut, C. Kocabas, S. Suzer, In-Situ XPS Monitoring and

Characterization of Electrochemically Prepared Au Nanoparticles in an Ionic Liquid, ACS Omega. 2 (2017) 478-486. doi:10.1021/acsomega.6b00456.

[18] A. Steel, S.W. Carr, M.W. Anderson, i __, (1996) 1829-1832.

[19] P. Mester, C. Robben, A.K. Witte, R. Kalb, M. Ehling-Schulz, P. Rossmanith, T. Grunert, FTIR Spectroscopy Suggests a Revised Mode of Action for the Cationic Side-Chain Effect of Ionic Liquids, ACS Comb. Sci. 21 (2019) 90-97. doi:10.1021/acscombsci.8b00141.

[20] D.W. Woodall, B. Wang, E.D. Inutan, S.B. Narayan, S. Trimpin, High-throughput characterization of small and large molecules using only a matrix and the vacuum of a mass spectrometer, Anal. Chem. 87 (2015) 4667-4674. doi:10.1021/ac504475x.

[21] Y. Chen, J.O. Iroh, Synthesis and characterization of polyimide/silica hybrid composites, Chem. Mater. 11 (1999) 1218-1222. doi:10.1021/cm980428l.

[22] V. Pascanu, A. Bermejo Gómez, C. Ayats, A.E. Platero-Prats, F. Carson, J. Su, Q. Yao, M.A. Pericàs, X. Zou, B. Martín-Matute, Double-supported silica-metal-organic framework palladium nanocatalyst for the aerobic oxidation of alcohols under batch and continuous flow regimes, ACS Catal. 5 (2015) 472-479. doi:10.1021/cs501573c.

[23] M. Kanezashi, K. Yada, T. Yoshioka, T. Tsuru, Design of Silica Networks for Development of Highly Permeable Hydrogen Separation Membranes with Hydrothermal Stability, (2009) 414-415.

[24] A. Mahtabani, I. Rytöluoto, R. Anyszka, X. He, E. Saarimäki, K. Lahti, M. Paajanen, W. Dierkes, A. Blume, On the Silica Surface Modification and Its Effect on Charge Trapping and Transport in PP-Based Dielectric Nanocomposites, ACS Appl. Polym. Mater. 2 (2020) 3148-3160. doi:10.1021/acsapm.0c00349.

[25] D. Montes, J. Henao, E.A. Taborda, J. Gallego, F.B. Cortés, C.A. Franco, Effect of Textural Properties and Surface Chemical Nature of Silica Nanoparticles from Different Silicon Sources on the Viscosity Reduction of Heavy Crude Oil, ACS Omega. 5 (2020) 5085-5097. doi:10.1021/acsomega.9b04041.

[26] G. Lubes, M. Goodarzi, Analysis of Volatile Compounds by Advanced Analytical Techniques and Multivariate Chemometrics, Chem. Rev. 117 (2017) 6399-6422. doi:10.1021/acs.chemrev.6b00698.

[27] C.Y. Li, J.H. Gao, J. Yi, X.G. Zhang, X.D. Cao, M. Meng, C. Wang, Y.P. Huang, S.J. Zhang, D.Y. Wu, C.L. Wu, J.H. Xu, Z.Q. Tian, J.F. Li, Plasmon-Enhanced Ultrasensitive Surface Analysis Using Ag Nanoantenna, Anal. Chem. 90 (2018) 2018-2022. doi:10.1021/acs.analchem.7b04113.

[28] L. Nguyen, F.F. Tao, Y. Tang, J. Dou, X.J. Bao, Understanding Catalyst Surfaces during Catalysis through Near Ambient Pressure X-ray Photoelectron Spectroscopy, Chem. Rev. 119 (2019) 6822-6905. doi:10.1021/acs.chemrev.8b00114.

[29] S.T. Abrahami, T. Hauffman, J.M.M. De Kok, J.M.C. Mol, H. Terryn, XPS Analysis of the Surface Chemistry and Interfacial Bonding of Barrier-Type Cr(VI)-Free Anodic Oxides, J. Phys. Chem. C. 119 (2015) 19967-19975. doi:10.1021/acs.jpcc.5b05958.

[30] D.R. Baer, K. Artyushkova, C. Richard Brundle, J.E. Castle, M.H. Engelhard, K.J. Gaskell, J.T. Grant, R.T. Haasch, M.R. Linford, C.J. Powell, A.G. Shard, P.M.A. Sherwood, V.S. Smentkowski, Practical guides for x-ray photoelectron spectroscopy: First steps in planning, conducting, and reporting

XPS measurements, J. Vac. Sci. Technol. A. 37 (2019) 031401. doi:10.1116/1.5065501.

[31] C. Blog, S. In, CMMP X-ray Photoelectron Spectroscopy, (2013) 2-4.

[32] F.A. Stevie, R. Garcia, J. Shallenberger, J.G. Newman, C.L. Donley, Sample handling, preparation and mounting for XPS and other surface analytical techniques, J. Vac. Sci. Technol. A. 38 (2020) 063202. doi:10.1116/6.0000421.

[33] FLUXANA, L. Peters, Sample Preparation: Making Pressed Pellets, (2019) 1-13.

[34] M. Mägi, E. Lippmaa, A. Samoson, G. Engelhardt, A.R. Grimmer, Solid-state high-resolution silicon-29 chemical shifts in silicates, J. Phys. Chem. 88 (1984) 1518-1522. doi:10.1021/j150652a015.

[35] D.M. Dawson, R.F. Moran, S.E. Ashbrook, An NMR Crystallographic Investigation of the Relationships between the Crystal Structure and 29Si Isotropic Chemical Shift in Silica Zeolites, J. Phys. Chem. C. 121 (2017) 15198-15210. doi:10.1021/acs.jpcc.7b03730.

[36] M.P. Conley, C. Copéret, C. Thieuleux, Mesostructured hybrid organic-silica materials: Ideal supports for well-defined heterogeneous organometallic catalysts, ACS Catal. 4 (2014) 1458-1469. doi:10.1021/cs500262t.

[37] D. Finlayson, C. Rinaldi, M.J. Baker, Is Infrared Spectroscopy Ready for the Clinic?, Anal. Chem. 91 (2019) 12117-12128. doi:10.1021/acs.analchem.9b02280.

[38] A.R. Brougb, C.P. Grey, C.M. Dobson, Paramagnetic Ions as Structural Probes in Solid-state NMR : Distance Measurements in Crystalline Lanthanide Acetates, (1993) 1318-1321.

[39] F.D. Doty, Solid State NMR Probe Design, Encycl. Magn. Reson. (2007) 1-21. doi:10.1002/9780470034590.emrstm1001.

[40] G. Althoff, Solid State NMR Hardware Probeheads, MAS Rotors, RF Filters Gerhard Althoff Bruker Biospin GmbH, (n.d.).

[41] G. Renner, P. Sauerbier, T.C. Schmidt, J. Schram, Robust Automatic Identification of Microplastics in Environmental Samples Using FTIR Microscopy, Anal. Chem. 91 (2019) 9656-9664. doi:10.1021/acs.analchem.9b01095.

[42] D. Ami, P. Mereghetti, A. Foli, M. Tasaki, P. Milani, M. Nuvolone, G. Palladini, G. Merlini, F. Lavatelli, A. Natalello, ATR-FTIR spectroscopy supported by multivariate analysis for the characterization of adipose tissue aspirates from patients affected by systemic amyloidosis, Anal. Chem. 91 (2019) 2894-2900..

[43] P. Bassan, J. Mellor, J. Shapiro, K.J. Williams, M.P. Lisanti, P. Gardner, Transmission FT-IR Chemical Imaging on Glass Substrates: Applications in Infrared Spectral Histopathology, (2014).

[44] D. Kuehl, P.R. Griffiths, Dual-Beam Fourier Transform Infrared Spectrometer, 50 (1978) 418-422.

[45] S. Preparation, A enuated Total Re fl ection (ATR) Mode-Advantages for FT-IR spectroscopy, Manual for operation, (2012) 1-11.

[46] J. Balajka, J. Pavelec, M. Komora, M. Schmid, U. Diebold, Apparatus for dosing liquid water in ultrahigh vacuum, 083906 (2018) 1-6.

[47] Fourier Transform Infradred Spectroscopy, SOP, 01890 (2011) 10-25.

Section 3

Techniques for Particle, Tissue and Cellular Separation

Chapter 4

Preparation of Tissues and Heterogeneous Cellular Samples for Single-Cell Analysis

E. Celeste Welch and Anubhav Tripathi

Abstract

While sample preparation techniques for the chemical and biochemical analysis of tissues are fairly well advanced, the preparation of complex, heterogenous samples for single-cell analysis can be difficult and challenging. Nevertheless, there is growing interest in preparing complex cellular samples, particularly tissues, for analysis via single-cell resolution techniques such as single-cell sequencing or flow cytometry. Recent microfluidic tissue dissociation approaches have helped to expedite the preparation of single cells from tissues through the use of optimized, controlled mechanical forces. Cell sorting and selective cellular recovery from heterogenous samples have also gained traction in biosensors, microfluidic systems, and other diagnostic devices. Together, these recent developments in tissue disaggregation and targeted cellular retrieval have contributed to the development of increasingly streamlined sample preparation workflows for single-cell analysis technologies, which minimize equipment requirements, enable lower processing times and costs, and pave the way for high-throughput, automated technologies. In this chapter, we survey recent developments and emerging trends in this field.

Keywords: tissue dissociation, cell sorting, microfluidics, diagnostics, devices

1. Introduction

The common conception of *in vitro* diagnostics is intertwined with the idea of a liquid starting sample - blood, saliva, urine, and other starting materials are often the candidates for study, allowing rapid determination of important details about a patient's health status by investigating metabolomic, proteomic and genomic markers of disease [1]. Tissue samples are comparatively less discussed, with much less research devoted to optimizing their sample preparation for analysis. Despite the comparative lack of research in this area, tissue specimens are ubiquitous diagnostic samples, particularly for cancer, where they are used to confirm and characterize almost every case of solid-tumor cancer (Mayo Clinic, 2020).

While liquid biopsy has seen much recent innovation as a potential cancer diagnostic, the paucity of relevant tumor cells in the collected liquids often fails to adequately represent the cellular heterogeneity present within cancer tissues. Consequently, while liquid biopsy, diagnostic imaging and other tests are often conducted for preliminary diagnosis of cancer, tissue biopsies and cancer tissue specimens remain the standard and are routinely needed in order to profile an individual's cancer and assess prognosis, metastasis, treatment options, and more. As such, millions of tissue biopsy procedures are conducted every year in the United States.

In the current practice, repeat tissue biopsies are often required and much of the biopsied materials end up being wasted. Sources have estimated that the United States spends $8 billion annually on unnecessary repeat biopsy procedures just for breast cancer [2]. This is often because bulk sequencing and histopathological analyses limit the full extent of investigation. There is a growing interest in applying Single-Cell Analysis (SCA) techniques to better understand tissue heterogeneity and to improve the efficiency of sample recovery in tissue diagnostics. SCA is an umbrella term for any tests that ascertain cellular characteristics at the individual cellular level. The most popular emerging SCA approach for cancer diagnostics is arguably Single-Cell Sequencing (SCS) of nucleic acids (DNA, RNA), which also encompasses techniques such as single-cellRNA sequencing (scRNAseq). This is a Next Generation Sequencing (NGS) approach to characterize the genomes or transcriptomes of individual cells.

While traditional approaches involve homogenizing whole tissue sections and extracting DNA from bulk tissues, SCS techniques and traditional genomic or transcriptomic approaches involve isolation of single cells for sequential high throughput analysis. Because of this, there is a growing interest in using SCS techniques for cancer diagnostics, as they are able to significantly reduce background noise and improve resolution [3]. However, a critical component which presently limits the clinical translational potential of the SCS workflow remains under-investigated: How does one get from a complex, heterogeneous section of tissue to a suspension of targeted cancer cells? We herein provide an overview of the different approaches whereby suspensions of single targeted cells can be prepared from complex tissues, with an emphasis on applications that target advanced SCS analysis (**Figure 1**).

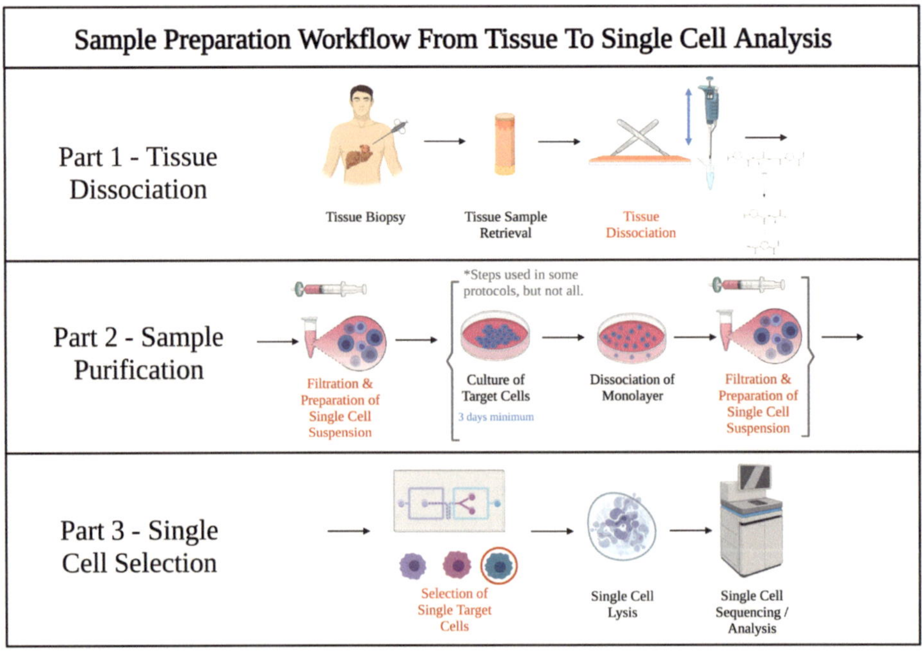

Figure 1.
Overview of the conventional sample preparation workflow from tissue to single cell sequencing. Topics in red will be discussed at length in this chapter.

DOI: http://dx.doi.org/10.5772/intechopen.100184

2. Tissue dissociation methods

2.1 Introduction to conventional tissue dissociation

We will begin by summarizing the conventional approaches currently used for tissue dissociation - the process of obtaining cellular suspensions from tissues. We will discuss recent innovations at the academic and commercial level and will conclude by discussing some possible future directions for innovations in this area.

Current methods for dissociating tissues into single cells rely on chemical dissociation techniques, mechanical dissociation techniques, or a combination of the two. The simplest benchtop preparation methods often consist of mincing tissues repeatedly with scalpels, followed by chemical treatment with proteases such as collagenase and other reagents that disrupt the extracellular matrix and cell–cell connections, often while agitating the solution. This is then followed by vortexing, centrifugation and other steps to aid in the purification of single-cell solutions for downstream analysis. In general, these protocols are sequential, with many prolonged and somewhat tedious manual steps, often resulting in a total processing time of an hour per clinically-sized biopsy sample.

The current standard of preparation techniques for tissue dissociation have been characterized extensively in the literature. Specific approaches has been developed for different tissue types, biopsy sizes, and downstream applications with the objective of producing a homogeneous single-cell suspension devoid of debris, aggregates, or off-target cells. A number of specific protocols can be found in resources such as the Worthington Tissue Dissociation Encyclopedia.

Early attempts to improve upon tissue dissociation protocols for particular tissue types of applications date back to the 1970s with investigation into different proteolytic enzymes and enzymatic adsorption at the cell surface [4]. Research has since

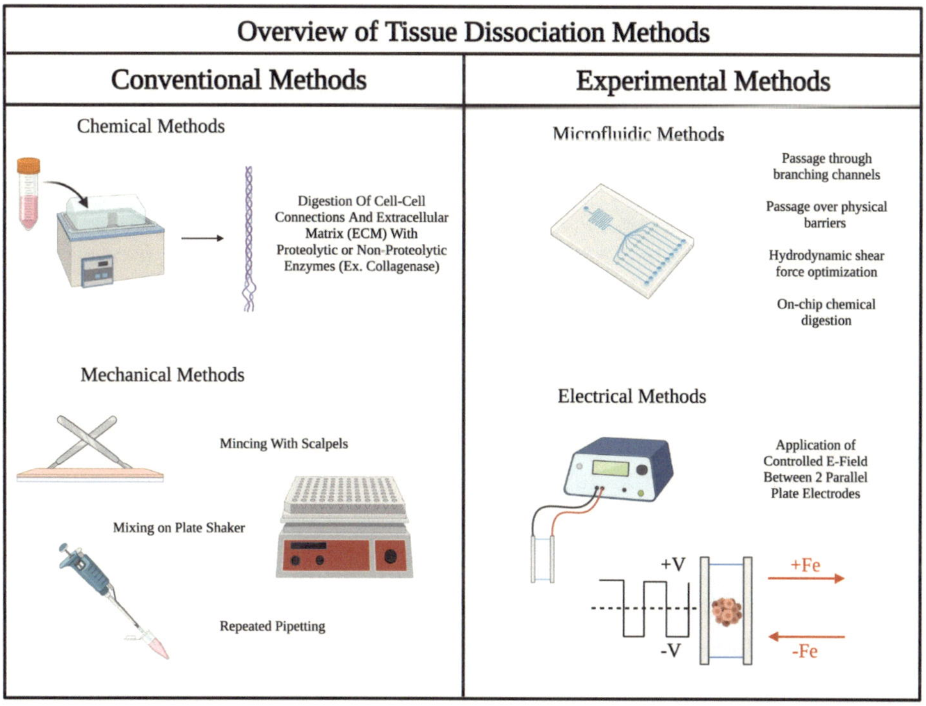

Figure 2.
Overview of various conventional and experimental tissue dissociation methods.

been focused in three main areas: (1) optimizing traditional chemical/mechanical benchtop protocols, (2) translating chemical/mechanical protocols into microfluidic systems, (3) translating chemical/mechanical protocols into commercial sample processing devices (**Figure 2**).

2.2 Optimization of benchtop chemical and mechanical tissue dissociation

Some research has been conducted on the subject of optimizing chemical and mechanical digestion of tissue. As mentioned, both chemical and mechanical dissociation protocols and outcomes are dependent on tissue type. An example of this is illustrated in the work of Robin et al., in which myogenic cells were isolated using collagenase D and dispase II with repeated pipetting, but skin fibroblasts were isolated by collagenase treatment and subsequent mincing [5].

Some research tests numerous enzymes and combinations to determine best chemical conditions for a given tissue. For example, in our recent work, we studied chemical processing treatments and then combined processing with simple mechanical agitation of samples in a 96-well plate on a mechanical plate shaker to determine optimal treatments for dissociation of single cells from bovine liver samples [6]. Physical and mathematical modeling was used to predict optimum mechanical mixing parameters.

While a great deal of research has been conducted in the area of chemical dissociation, comparatively less work has been done on improving benchtop mechanical dissociation. Tissue grinder approaches are a common feature in the literature but are generally not indicated for downstream single-cell analysis due to their tendency to disrupt the cell membrane. Recent work by Scheuermann et al., however, used a tissue grinding benchtop preparation workflow created with downstream SCA in mind [7].

2.3 Microfluidic tissue dissociation

Various microfluidic approaches to the dissociation of tissue and cellular aggregates for recovery of single cells have been investigated by researchers in recent years. Many of the investigated approaches consist of microfluidic chips or processing workflows that use either a purely chemical or purely mechanical approach, but there have also been recent advances which combine chemical and mechanical processing in a single device.

Purely chemical microfluidic approaches to tissue dissociation are present in the literature but are somewhat less explored. An example of a chemically-focused tissue maintenance and dissociation microfluidic device is that created by Hattersley et al., which was used to maintain cellular viability in liver tissue biopsies over 70 hours [8]. In addition to this, the researchers performed an on-chip collagenase digestion.

Other mechanical and chemo-mechanical approaches involve applying force against a physical barrier or using hydrodynamic flow-based disruption to dissociate cellular aggregates and tissues. More recently, both of these methods have been combined into single microfluidic systems for improved results.

2.3.1 Applied force against a physical barrier

A common approach in microfluidic tissue dissociation devices involves the passage of cell aggregates through or over objects to physically disrupt cell–cell bonds. Lin et al., created an enzyme-free microfluidic chip for neurosphere dissociation which used micropillars with 20 μm gaps [9]. When a driving force was applied via a syringe pump, the spheroids were driven through the gaps, separating them into their constituent single cells. Another similar approach is the Biogrid, a silica-knife microfluidic device created by Wallman et al. for enzyme-free dissociation of stem cell aggregates [10]. The Biogrid knives had edge thicknesses of 30 μm and grid spacing of 200 μm. Another example of physical-barriers in microfluidic tissue dissociation is the microfluidic filter device reported by Qiu et al., which consisted of 2 nylon mesh membranes with 50 and 15 μm mesh sizes [11]. The mesh can be used to both elute single cells and retain cellular aggregates for further processing.

Physical-object based disruption can be an effective method to separate cellular aggregates into smaller aggregates and single cells, and (in the case of mesh) to selectively retain or elute cells and aggregates based on size. However, for these methods to work, there must be an applied force, usually in the form of a syringe pump-generated pressure-driven flow. Additionally, sample loss and cell death can occur with processing, as cells collide with, deform, and stick to or are ruptured by the objects in their path. Sample constituents, such as cell free DNA and cellular debris, can also lead to device fouling, causing significant clogging of narrow apertures through which cells must pass.

2.3.2 Flow-based disruption

Another method for cellular aggregate disruption is passage through a mechanically-optimized dissociating microchannel. Examples often include concepts such as hierarchical or branching microchannels, in which subsequent channels are half the width of prior channels. This principle is used in a set of two papers by Qiu et al., in which a network of branching channels is created and then optimized with shark-fin geometry in order to increase cellular recovery of single cells from aggregates passing through the chip [12]. The channels also used repeated channel expansions and constrictions in order to produce hydrodynamic fluid jets [13], which imparted forces on cellular aggregates, aiding in dissociation into single cells.

Other, previously utilized approaches include placing tissue specimens within a confined area on a chip and applying fluid jets to the tissue core. This concept is present in another paper by Qiu et al., in which the researchers use precision-flows to create a hydrodynamic mincing device [14]. The device accommodates tissue specimens up to 1 cm in length and 1 mm in diameter, focusing shear forces at distinct locations to improve contact between areas of the tissue and enzymes that aid in the disruption of cell–cell adhesion.

2.3.3 Integrated microfluidic tissue dissociation devices

In 2021, a notable work by Lombardo et al. integrated three different tissue processing techniques on a single microfluidic chip – enzymatic digestion, hydrodynamic shear force, and mesh filtration - arguably representing the first chip of this kind [15]. This indicates a progression towards a "Next Generation" of microfluidic tissue dissociation devices. Their system combines tissue digestion, disaggregation, and filtration steps, and was tested on an array of kidney and mammary tissues. Using this technique, 2.5-fold greater recovery of single cells was obtained than with traditional techniques. They also demonstrated recovery of target cell type numbers at particular timepoints during device processing and demonstrated the utility of their device in a single-cell sequencing study. Progress in this area is likely to gain traction in the coming years.

2.4 Commercial tissue dissociation devices

Companies such as Miltenyi Biotec and S2Genomics have attempted to address the need for obtaining single cells from tissues by creating products that automate many of these functions, reducing manual labor and simplifying tissue dissociation workflows (e.g. the GentleMACS Dissociator). Miltenyi Biotec has made progress in addressing limitations imposed by differences among tissue sources and organs by selling dissociation kits specifically designed for organism (e.g. mouse vs. human) and organ (e.g. small intestine, lung, tumor, brain). Despite this progress, there remains a significant need for improving the size, cost, complexity, and performance metrics of these devices in order to enable improved cancer diagnostics via SCS. Key limitations of the GentleMACS Dissociation system, for example, are high costs in excess of $500 USD for only 25–50 preparations.

2.5 Electrical tissue dissociation

Although it has only been recently characterized, applied oscillating square wave voltages have been shown to aid in the dissociation of complex tissues into their component single cells [16]. Excellent cellular dissociation was observed with 1 kHz electrical conditions in a matter of <5 minutes, significantly faster than conventional chemo-mechanical approaches. The exact mechanism of the observed electric field induced tissue dissociation is not completely understood at this time but may be related to known effects of electric fields on membrane potential, ion transport, cellular movement, and proliferation.

A key finding of this work is the importance of oscillation frequency on tissue dissociation. Higher oscillations in the MHz range have previously been used to decrease heating and electroosmotic effects in electrophoretic applications [17], and previous research has shown that cells can reorient themselves in such electric fields in a manner dependent on the orientation and shape of the cell, as well as the electric field properties [17]. Presumably, the significantly lower dielectric constant of the cell membrane, and higher conductivity than the medium of suspension or cytoplasm can lead to charge accumulation at the membrane, and induced dipole moments that can generate shear stresses within the tissue, leading to alignment, rotation, and stretching behaviors. It is also thought that application of electric field induces a torque on the membrane glycocalyx, which is analogous to an externally applied mechanical force, and results in movement [18].

2.6 Regional dissociation methods

Within tissues, different regions of interest exist with different cellular properties. These regions can often be identified by microscopy or other methods but may be difficult to pick out using conventional heterogeneous sequencing approaches. Approaches such as manually excising an area of interest in a tissue sample for later analysis can still include numerous off-target cells.

Selective dissociation of cells from specific target regions of interest has been an area that has fascinated scientists since the 1980s. In the first work of its kind, Freyer and Sutherland selectively dissociated and characterized cells from various regions of tumor spheroids by applying a dilute trypsin solution at 18 to 30 degree angles in custom dishes [19]. Freyer and Schor later characterized the dissociation of cells from different spheroidal regions using an automated device that exposed outer spheroidal layers to 0.25% trypsin [20].

More recent advances in regional dissociation and regional tissue interrogation have also been made. The most notable technique in the field of region of interest analysis is arguably laser capture microdissection, a technique that is integrated with common optical microscopy infrastructure and used with conventional tissue-on-slide imaging. This technique uses optical interrogation to manually or automatically select tissue regions of interest and separate them from the surrounding tissue, placing targeted cells in a tube for subsequent molecular analysis. Other

ROI techniques such as Computer-Aided Laser Dissection can incorporate machine learning and other informatics techniques into the workflow for automated computer recognition of regions of interest for subsequent selection.

2.6.1 Laser capture microdissection

The laser capture microdissection (LCM) process dates back to high-profile works published in *Science* in 1996 and 1997 by Emmert-Buck et al., and Bonner et al., respectively [21, 22]. The IT-based workflow begins with an initial microscopic visualization, followed by selection of cells of interest on a conventional microscopy slide. A region of interest (ROI) is then manually selected in conventional LCM workflows. After ROI selection, a cap covered with transfer film is lowered onto the tissue surface. A laser pulse is then activated, which in turn activates the transfer film, causing the cells in the selected area to adhere to the film on the cap. In the UV-based LCM technique, the laser is able to cut around the cell itself, bypassing the plastic adhesion process [23]. The selected cells can then be placed into a tube for molecular analysis and other downstream tests. The technique can also be used as an alternative to histological staining in proteomic sample analysis [24]. To date, combination of dissociation of the selected tissue for SCA has received only scant attention, but further research in this area over the next few years would represent a logical progression of existing microdissection technology (**Figure 3**).

2.6.2 Computer-aided laser dissection and machine learning

In the years since the creation of the LCM technique, many researchers have recognized a key limitation in the lack of automation of LCM: the requirement for manual selection. This results in low-throughput, making processing large amounts

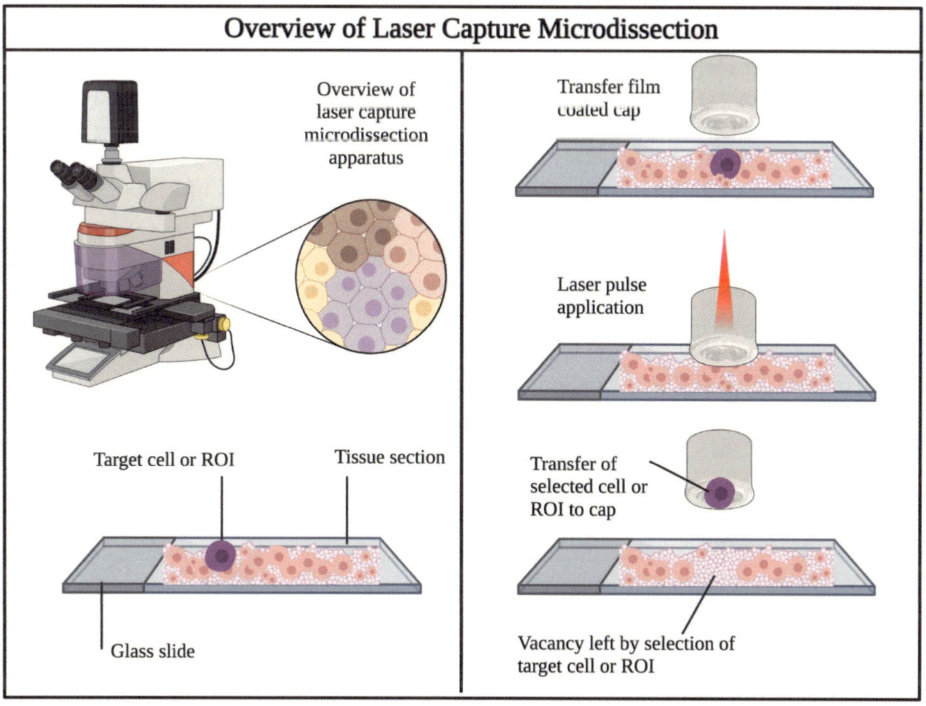

Figure 3.
Overview of laser capture microdissection.

of tissue samples a difficult task. A proposed solution lies in the Computer-Aided Laser Dissection (CALD) technique, which uses algorithms to recognize areas of interest and oversee dissection in an automated or semi-automated manner.

Statistical learning methods have long been proposed as a tool for region of interest detection in cancer tissue specifically, but have also been widely used in the interpretation of X-rays and other biomedical image data for diagnostic purposes [25, 26]. In recent work by Hipp et al., authors compare use cases of integrating image analysis tools into the LCM workflow [27]. They discussed possibilities of integrating existing software like ImageJ, as well as techniques such as spatially invariant vector quantization and probabilistic pairwise Markov models. Recently, Ren et al. have translated a machine learning technique to achieve automatic partitioning of selected regions of interest [28]. They used a K-means clustering algorithm with a fuzzy c-means clustering algorithm to discern ROI, and automatically selected clustering results. It is likely that this particular area will experience significant growth in the coming years.

A major limitation of computer vision and machine learning techniques for region of interest detection in tissue samples is the need for algorithm training. Additionally, different algorithms must be used to detect regions of interest in different tissue types, as regions of interest can have distinct visual morphologies across tissues. Furthermore, these techniques may be poorly equipped to deal with irregular appearances of regions of interest, which could pose a hurdle towards their translation in cancer diagnostics.

3. Single-cell suspension purification methods

After tissue dissociation into cellular suspensions, workflows must be created that enable selection of single-cell populations for downstream analysis. A critical step in getting purified single-cell suspension lies in the discernment of single cells from large cellular aggregates and debris from the extracellular matrix and lysed cells.

We herein provide an overview of methods for removal of cellular clumps and debris, a sample preparation step which has become known as purification. These procedures are unique from cell sorting and manipulation methods due to the nature of their position in the pipeline from tissue to SCS. While cell sorting and manipulation methods are occasionally equipped to separate single cells from doublets, for example, they generally cannot be used without an intermediate purification step when analyzing dissociated tissue samples.

These sample purification processes are distinct from target cell selection methods as they are not focused on separating rare target cells from heterogeneous suspensions but are instead focused on simply obtaining purified single-cell suspensions themselves. While it has been shown that techniques such as flow cytometry facilitated FACS can be used directly from dissociated tissue samples, this section will focus on conventional methods of purification. Methods that will be discussed here are filtration and chemical debris removal (**Figure 4**).

3.1 Filtration

Filtration techniques are currently considered to be the gold standard for the sample purification step. Commercially available mesh filters can be purchased and incorporated into the benchtop preparation workflow when processing tissue into single-cell suspensions for SCS. These commercial filters, created by companies such as Miltenyi Biotec, are often sold as "caps" that fit directly into centrifuge tubes. The tissue dissociate is simply poured over the cap, and filtrate passes through into the bottom of the tube. While this represents an example of

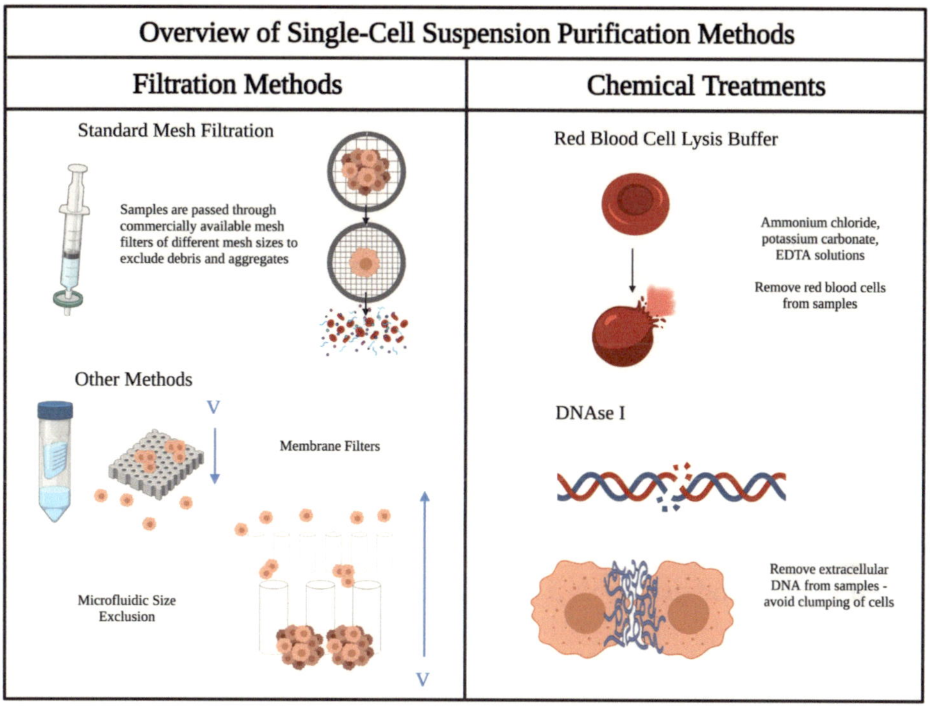

Figure 4.
Overview of single-cell suspension purification methods.

conventional filtration, other filtration methods can also be used. Different filter types include membrane filters, post filters, and weir filters.

Membrane filters contain porous structures that can be used to trap larger cells, such as tissue cells, while excluding smaller cells, such as red blood cells. Pillar based filters and weir filters all consist of obstacles about which cells must navigate. Weir filters are characterized by a singular obstruction nearly closing off the channel, permitting only particles small enough to navigate the gap through, while pillar filters consist of numerous evenly spaced pillars. These systems are all prone to clogging with debris and clumps of tissue, especially at high cellular densities, although systems which use dynamic filtration instead of barrier filtration can clog less. Forward and reverse flow can be used to unclog filters if they are incorporated into a microfluidic environment, but clogging still poses an issue, nonetheless.

Other filtration techniques of interest include cross-flow filtration and size exclusion filtration. Cross-flow filtration is a filtering technique that has been used in the purification of animal cells since the 1990s [29]. It offers a prospective solution to one of the most ubiquitous issues of all filter-based cellular purification systems, clogging, by using sieve-like sorting. Size exclusion filtration can be created with progressively decreasing filter mesh sizes but is occasionally used to refer to pillar-based filters containing tiered post arrangements with decreasing gaps. These systems can be translated into microfluidic chip formats in order to isolate differently sized particles in different regions of the chip [11, 30].

3.2 Chemical methods for debris removal

In addition to mechanical filtration of dissociated cellular suspensions, chemical methods for debris removal are also highly utilized. Common methods include removal of red blood cells, and removal of extracellular DNA.

In the SCS workflow, non-nucleated red blood cells are often excluded from analysis by chemical treatment with products such as red blood cell lysis buffers that are commercially available for purchase. These buffers can contain ammonium chloride, potassium carbonate, and EDTA. The solution is incubated on pelleted dissociated tissue for 4–5 minutes at room temperature, and then deactivated by adding PBS, and subsequently removed, according to BestProtocols guidelines provided by ThermoFisher Scientific.

Extracellular DNA removal is also a popular purification step for some SCA protocols. During the tissue dissociation process, extracellular DNA is commonly released. The DNA sticks to the cells, causing them to form aggregates. This can cause aggregation on filters, which can lead to sample loss and clogging. DNAse I solutions are a proposed answer to this problem. They are recommended to be applied to solutions at a concentration of 100 μg/mL, according to BestProtocols. An issue with DNAse I use is that it is inhibited by actin release from dead cells. It also must be removed from the sample for successful downstream analysis with SCS.

4. Cell sorting and manipulation methods

After the sample purification step, many single-cell analysis workflows require additional processing steps to eliminate any remaining doublets or off-target cells. Therefore, some SCS workflows are concerned not only with the selection of individual single cells, but also the retrieval of specific subsets of cells from heterogeneous cellular populations. Examples include selecting for cancerous cells from a predominantly healthy cell population or selecting for cells with rare genetic mutations. Sorted suspensions of cells of interest are then able to enter the remaining part of the SCS workflow, consisting of lysis and analysis steps.

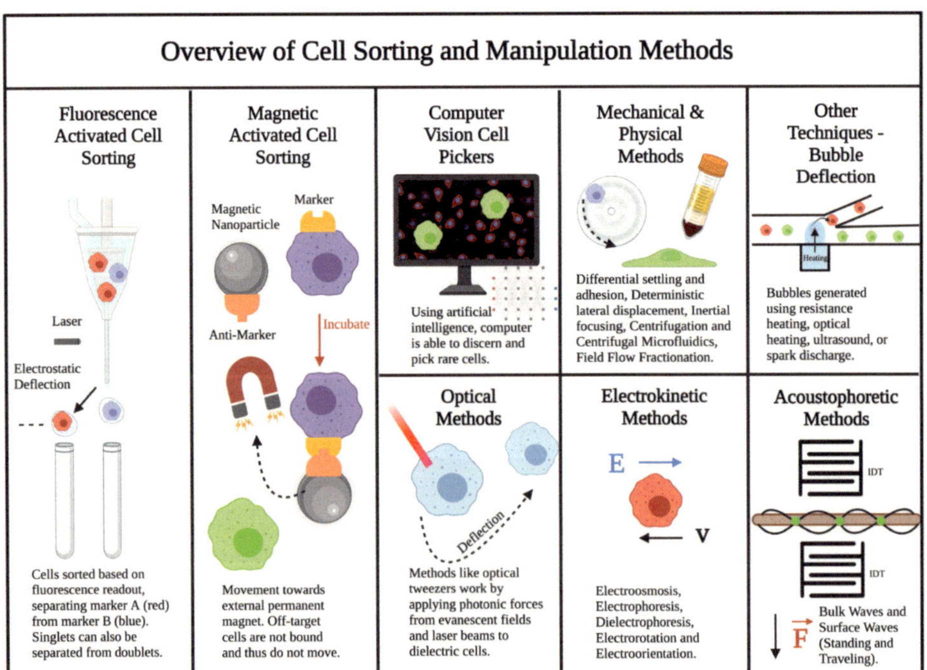

Figure 5.
Overview of cell sorting and manipulation methods.

Methods that will be discussed in this section cover techniques for cell sorting and manipulation, many of which have relevance to target cell selection. Topics to be discussed include fluorescence activated cell and droplet sorting, magnetic activated cell sorting and magnetophoresis, computer vision and cell picking techniques, optical techniques such as optical tweezers and focused beams, mechanical and physical cell sorting, electrokinetic and acoustophoretic cell sorting (**Figure 5**).

4.1 Fluorescence activated cell sorting

4.1.1 FACS history and commercial flow cytometry

Fluorescence Activated Cell Sorting (FACS) is a ubiquitous technique created in the late 1960s by "Bonner, Sweet, Hulett, Herzenberg, and others" [31, 32]. The technique built upon previous cell sorting developments in flow cytometry and is currently used widely by researchers in order to obtain cellular suspensions of target cells.

In the conventional FACS process using a standard flow cytometer, a cellular suspension sample is injected through a fluid sheath into a flowing stream in laminar flow conditions. The cellular suspension is previously treated with a fluorescent probe or antibody, which can be either a general probe, or a probe specific for a particular cell type or characteristic of interest.

A laser illuminates the stream directly before droplet formation, and fluorescence emission scatter information is transmitted to a detector. Immediately afterwards, aerosolized droplets are produced by vibration of the instrument, and contain either a single-cell or an absence of cells, as verified using the recorded fluorescence signal information. The droplets are charged based on cellular contents and are then electrostatically deflected into containers based on charge. Uncharged droplets maintain their original course.

Since the invention of FACS, numerous technological improvements have been developed, such as incorporation of multiple fluorescent lasers and detection capabilities within commercially available flow cytometry instruments [32]. FACS techniques are constantly evolving, and have also notably been translated outside of the flow cytometry workflow and into microfabricated microfluidic chip systems over the past two decades [33, 34]. This translation can circumvent limitations of the traditional aerosolization process while still maintaining the general principles of FACS such as real-time sample classification and binning. While traditional FACS is an open-system and lacks sterilization, microfluidic FACS chips are sterile and single use.

Other persistent limitations of FACS are relatively high cell number requirements ($>10^6$ cells) and reagent consumption, high cost of flow cytometry instruments, and the potential for cross-contamination, and clogging, especially when analyzing dissociated tissue samples [35].

4.1.2 Fluorescence activated droplet sorting

Fluorescence Activated Droplet Sorting (FADS) is a microfluidic fluorescence-activated sorting technique which uses many of the same principles of FACS [36]. FADS was first described by Baret et al. in 2009 as a method that combined the benefits of FACS with those of microtiter-screening. The process involves encapsulation of single cells within emulsion droplets, which are then dielectrophoretically sorted as in FACS. In Baret et al.'s initial work, sorting rates of up to 2,000 droplets/second were obtained with a false positive rate of $<1/10^4$ droplets. Limitations were seen in co-encapsulation of fluorogenic substrate with cells, not in sorting. Errors could be minimized at low sorting densities (<10^5 cells), which are unable to be run using conventional FACS flow cytometry [37]. Ever since this initial work, FADS has been used in numerous single-cell applications, and other applications, such as synthetic biology. Newer systems have focused on increasing throughput, and multiplexing droplet sorting capabilities [38].

4.2 Magnetic activated cell sorting and magnetophoresis

Magnetic Activated Cell Sorting (MACS), or magnetic separation of target cells is one of the most commonly used separation techniques for cellular selection. In this technique, nanosized superparamagnetic beads are bound to a particular antibody or other recognition element for a specific cellular target of interest. Specifically bound cells are then removed from solution using a magnet, followed by washing of the beads and liberation of the bound cells. Common cellular-recognition elements include the Epithelial Cell Adhesion Molecule (EPCAM) for separation of CTCs from whole blood, such as is used in the commercially available Veridex Cellsearch system and other applications [39]. Other approaches can use separation via specifically engineered indirect interactions, such as biotin/streptavidin coupling.

Once nanoparticles are co-incubated with cellular solutions, they can bind to target cells through antibody/antibody, antibody/antigen, or other interactions. The bound target cells can then be separated from unbound off-target cells and collected by moving them to a different location within a microfluidic chip or tube using an external permanent magnet, via a process known as "magnetophoresis". Electromagnetic coils can also be used to enact magnetophoresis, although this is a less common approach. Researchers such as Pirozzi et al. have used this technique in combination with microfluidic oil/water immiscible phase filtration in order to separate cancer cells from various sample matrices, including unprocessed whole blood.

MACS principles have been translated into commercially available systems. The first commercially available MACS system was created by Miltenyi in 1990 [40]. CTC enrichment of 10^5 fold was obtained using high magnetic gradient column MACS, but purification was found to be lower than would be clinically applicable in CTC work [35, 41]. This has since been improved using magnetic sweeping and other techniques.

Much work on MACS is focused on CTC isolation from blood using markers, a much easier problem to solve than isolation of target cellular populations from dissociated tissue samples. However, some work has been conducted on stem cell isolation of solid-tissue originated cells [42]. Other work has been conducted on MACS separation from tumor tissue, but there has been little progress in this area as MACS separation is not a label-free technique, and relies on marker-based separation of specific target cell types [43]. As such, MACS is not considered to be an appropriate technique for understanding the true extent of tissue heterogeneity but can be used within the processing workflow to isolate particular cell types of clinical interest with known unique extracellular markers.

4.3 Computer vision and robotic cell picking techniques

Automated cell picking instruments are recent commercially translated instruments which can take various forms. Common themes in cell pickers are the use of optical imaging and computer vision to distinguish cells of interest. Certain cell pickers, such as the Shimadzu Cell Picker, are focused on picking cell colonies off culture plates for use in cell line construction and cloning applications. Other systems are designed specifically for isolating rare cells from liquid biopsies, such as the CellCelector instrument, an Automated Lab Solutions product. Many are designed specifically with downstream single-cell genetic analysis in mind.

Cell pickers can use diverse methods to select cells of interest, such as image recognition of fluorescent cells, distinct cellular morphologies, and more. After selection occurs, cells must be physically "picked", or removed. Most cell picking devices work by using a mechanical micro-manipulation robotic picking system. For example, the Shimadzu Cell Picker works by simply lowering a pipette to a particular location, as ascertained using the automated picking system. The CellCelector is equipped with different picking modules using capillaries, metal scraping tips, or plastic tips. In addition to translation into commercially available instruments, computer vision and robotic sorting have also been incorporated into microfluidic cell sorting [44]. Similarly to LCM, robotic cell picking techniques are often lower-throughput in comparison to other methods (e.g. FACS), despite automation. They are also frequently performed in open-air, posing a contamination risk.

4.4 Optical techniques

Optical techniques can be used in order to physically trap cells in suspension using elements such as focused laser beams, fields, and more. Some commonly used optical sorting techniques include optical tweezers and focused optical beams. The optical tweezers technique is arguably the most well-characterized technique and can be used to noninvasively move and separate dielectric particles in solution, such as cells, in a label free manner. Optical beams use radiation forces to achieve cell movement, and can also use optical radiation for other applications, such as bubble-induced cellular deflection.

4.4.1 Optical tweezers

Optical tweezers, invented by Ashkin et al., in 1986, are used in the noninvasive label-free separation of cells [45]. They work by applying photonic forces from evanescent fields and laser beams. Depending on properties of the refractive indices, cells and other dielectric particles are acted on by forces that can hold them in place, or move them to a new location [46, 47]. Optical tweezers are highly sensitive instruments that enable applied force resolution in the piconewton range, with sub-nanometer spatial resolution. Thus, these are highly useful tools in the microfluidic separation of individual target cells.

In addition to their usefulness in sorting, optical tweezers have been applied to the interrogation of numerous biological questions from assessing cellular responses over time to analyzing cellular biomechanics [46, 48]. While they have been applied to the separation of embryonic stem cells and tissue engineering, they have yet to be comprehensively utilized within the tissue to single-cell workflow [46, 49, 50]. Robotics technology has also been integrated with optical tweezers in recent years [51–53].

4.4.2 Optoelectronic tweezers

Optoelectronic tweezers (OETs) combine optical manipulation with electronic capabilities, specifically electrode-based DEP, to create a promising label-free technique that is frequently higher-throughput than optical tweezers, and a potentially promising area for SCS sample preparation. Optical images are projected onto a surface, creating temporary visual electrodes, which are used to locally apply an electric field, producing a dielectric force which can be exerted on the cells. In the event that the cell is less polarizable than the liquid, such as with cells in culture media, the pattern will repel it, which can be used to confine a cell to a specific location while fluid and other materials flow freely [54]. In addition to OETs potential for trapping, cells can also be transported through the chip and sorted via dielectrophoretic forces via the creation of light-defined channels and use of size selective sorting, respectively [55].

The translation of optoelectronic tweezers is somewhat limited by high required optical intensities, as is also a key limitation of standard optical tweezers [56]. However, light intensities have been used for cell separation in OET devices that are 100,000-fold lower than optical tweezers [57]. Other advances, such as combination with microfluidics and electric field re-orientation, have been used to increase the potential for OET translation. OETs have been used to manipulate mammalian cells in a variety of ways, from replicating the tumor microenvironment on a microfluidic chip [58] to their combination with scRNAseq preparatory microrobotics [59].

4.4.3 Focused optical beams

Focused optical beams can be used in order to apply radiation force to individual cells in a microfluidic environment, resulting in precisely controlled cellular manipulation, similar to that observed in the optical tweezers technique [60]. Cells can be deflected into an alternative channel at a junction with the influence of a laser pulse [61].

Focused optical beams such as lasers can also be used in cell sorting in an additional way. Localized heating from laser light can result in bubble formation, which can in turn produce mechanical deflection of target cells [62]. Optical scalpel techniques can also be used in order to assess different cellular components, but these techniques are not in-line with the expected workflow for downstream single-cell analysis.

4.5 Label-free sorting by mechanical and other physical properties

Other methods can sort cells based on their physical properties such as density, size, and settling behavior. Techniques that will be discussed here include settling and adhesion-based techniques, deterministic lateral displacement, inertial focusing, centrifugal microfluidics, and field flow fractionation.

4.5.1 Settling and adhesion

Different cell types settle and adhere differently in *in vitro* environments, due to a combination of material interaction and mechanical properties. Bailey-Hytholt et al. recently demonstrated effective rare cell selection via different settling behavior of cells in polystyrene wells [63]. This approach was used in a prenatal testing workflow for the purpose of separating rare extravillous trophoblasts from infants *in utero* from maternal cervical cells in cervical swabs. Bailey-Hytholt also notably investigated trophoblast enrichment via differences in surface adhesion on an inclined plane [64]. The principles of differential adhesion have also been applied to embryonic cell sorting [65].

Differential surface adhesion can also be artificially engineered by coating a patterned surface on a microfluidic chip with a particular marker or ligand. Stott et al. were the first to take this approach to cellular isolation in 2010 using a Herringbone chip coated with EpCAM to detect CTCs in the blood [66]. EpCAM has also been used in other more recent approaches [67] and EpCAM-free approaches to cancer cell isolation have also been investigated by looking at specific peptides of interest for particular cancer types [68]. However, these approaches are not label-free, which significantly limits their translational potential into the tissue to SCS NGS workflow.

4.5.2 Deterministic lateral displacement

Deterministic lateral displacement (DLD) was initially used for particle separation in 2004 in Huang et al.'s notable work [69]. Since then, it has been widely used for cell separation, as characterized in detail elsewhere [70, 71]. Deterministic lateral displacement structures have been used in the label-free separation of large cells for tissue engineering [72], and cancer cells from diluted whole blood samples [73]. This method uses different size-dependent hydrodynamic forces to separate cancer cells from red blood cells and other cells as they pass through a channel equipped with numerous posts through which the cells must navigate. Various works have also shown that triangular posts can help to optimize results [73, 74].

A similar approach to DLD is deterministic cell rolling. Cell rolling devices consist of ridged microfluidic channels, which, like DLD, enter the flowpath of cells, resulting in mechanical deflection. The surface of the channel can be modified with a target of interest, such as P-selectin, which can briefly bind target cells, deflecting them to a different location within the chip [75].

4.5.3 Inertial focusing

Inertial focusing is another technique frequently used to sort cells in microfluidic chips by inducing movement across laminar flow streamlines. Continuous inertial focusing, ordering, and separation of particles was demonstrated by Di Carlo et al. in 2007, and has since been applied extensively in microfluidic cell separation [76]. Di Carlo et al. demonstrated that particles in laminar flow conditions can migrate across streamlines in a "continuous, predictable, and accurate manner". The migration is caused by particle lift forces in conditions where inertial flow is significant. Boundary effects at microchannel walls can create this lifting phenomenon. Particular channel geometries create inertial forces that result in a bias towards certain equilibrium positions, a phenomenon which can be used for ordering of cells within a given microfluidic chip. Inertial focusing has been used, for example, in spiral microchannels and serpentine channels [77–82]. Inertial effects can also be combined with other hydrodynamic techniques, or flow fractionation techniques as well, such as pinched flow fractionation.

4.5.4 Centrifugation and centrifugal microfluidics

Centrifugation techniques are some of the oldest methods historically used in the separation of various blood components, such as nucleated cells and red blood cells. This same principle has been used in the sorting of CTCs from other cells in blood and bone marrow samples [83]. While developments using standard tube-based centrifugation methods have been minimal in recent years, centrifugal microfluidics has spawned a new field of target cell separation research.

Centrifugal microfluidics techniques have been investigated for over 40 years [84]. Techniques have been created using recycled or repurposed compact disks or disk-shaped polymer-based materials, which are rotated around a conventional or custom centrifuge to separate rare cells from other sample components. These systems are referred to as Lab-on-a-disk or Lab-on-a-CD platforms [84]. Snider et al. recently created one such system for application in the separation of single cancer cells from dissociated biopsy cores [85]. They were able to observe up to 80% capture efficiencies when using 2,000 to 15,000 cell input samples. While some centrifugal microfluidics applications perform chemical and mechanical cell lysis on chip, this system preserved cells in-tact for visual microscopic analysis. Centrifugal microfluidics can be integrated with geometric, magnetophoretic, and dielectrophoretic separation approaches [86].

4.5.5 Field flow fractionation

Field flow fractionation (FFF) and, in particular, sedimentation field flow fractionation (SdFFF) is a label-free method that can be used to separate various cell types. While FFF is primarily utilized in the separation of subcellular particles, macromolecules and nanoparticles, SdFFF has been used in the separation of stem cells [87]. For example, SdFFF has been used in the isolation of cancer stem cells from colorectal cancer cell lines [88]. Pinched flow fractionation (PFF) is a similar technique, but it is mainly used in separation of blood components and non-mammalian cells [89, 90]. FFF techniques have also been combined with other techniques, such as dielectrophoresis [91, 92].

4.6 Label-free sorting by electrokinetic properties

With the exception of electrical tissue dissociation and FACS, electrokinetic methods for cell separation and recovery have heretofore not been discussed in this chapter. Electrokinetics are physical effects resulting from applied electric fields that produce physical movement of cells and other particles [93]. While our discussion of FACS mentioned electrostatic sorting of charged droplets, in this section, we will discuss electroosmosis, electrophoresis, and dielectrophoresis within the context of cellular separation from dissociated tissue samples. We will also briefly discuss electrorotation and electroorientation phenomena (**Figure 6**).

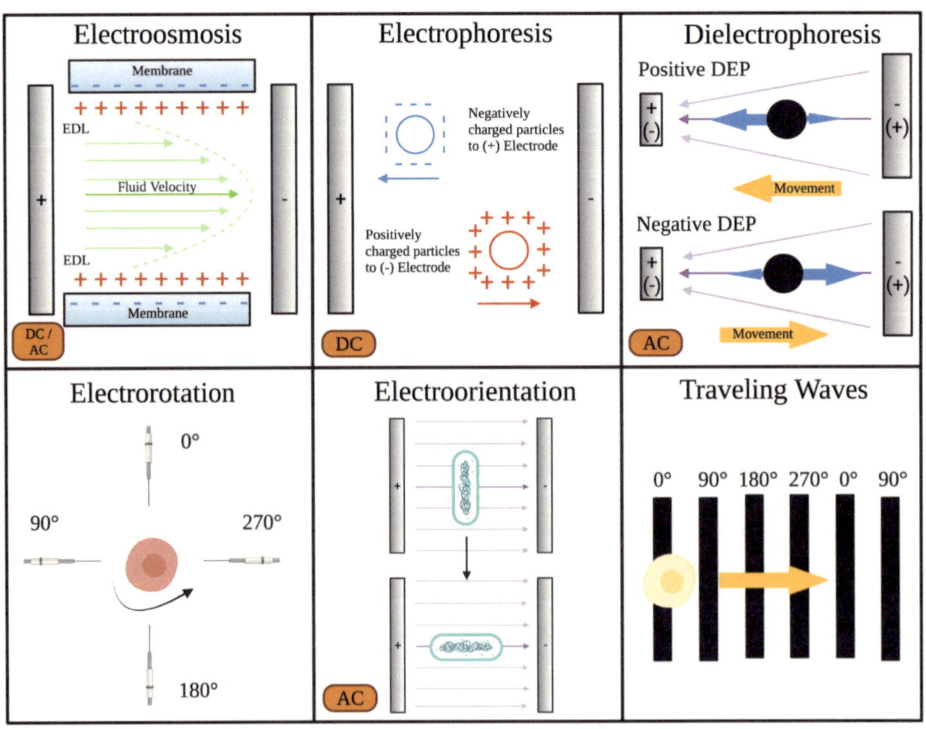

Figure 6.
Overview of electrokinetic phenomena used in the manipulation of cells.

4.6.1 Electroosmotic flow

Electroosmotic flow moves fluids and the particles suspended within them by induced migration of solvated ions [94]. Therefore, electroosmosis is characterized by fluid movement, as opposed to particle movement within a fluid, the mechanism at play in electrophoresis and dielectrophoresis. Electroosmosis has also been used in combination with FACS in order to sort fluorescent cells. Electroosmotic flow has also been used to create microfluidic on-demand particle separation by inducing a vortex on the conductive surface to direct particles, such as cells, towards an outlet. This phenomenon is referred to as controlled electroosmotic micro-vortexes or fluid pumps [95, 96]. Alipanah et al. have recently used induced charged electroosmotic flow and magnetic fields in conjunction, and managed to use flow vortexes to enable the magnetic field to overcome particle drag forces [97]. The effect can be tunable, as demonstrated by Yan et al., by adjusting modulating electrode and insulating post thickness, which correspondingly increases flux within the channel [95].

A major limitation of certain electroosmotic protocols lies in the use of DC electric fields. DC fields can induce electrolysis of water, bubbling, and production of reactive oxygen species, which can decrease viability of cells and introduce turbulence. In order to overcome these limitations, AC electroosmosis has gained some attention. For example, Puttaswamy et al. used AC electroosmosis to avoid these negative effects of DC field application [98].

4.6.2 Electrophoresis

In electrophoresis, particles suspended within a solution move towards an electrode with opposite charge in a DC field with a force proportional to their surface charge. Cells have a slightly negative charge, resulting in their migration towards the positively charged electrode. While electrophoresis is generally considered to be less efficient in the sorting of cells when compared to other processes such as dielectrophoresis, it has still been applied in the separation of mammalian cells [99]. One approach to mammalian cell separation via electrophoresis is with capillary electrophoresis, which can be coupled with an inkjet printing system. This approach was used to separate HUVEC, HepG2 and Caco-2 cell lines in a drop-on-demand technique by Zhang et al. [100]. Water-in-oil continuous flow techniques with encapsulated cells can also be used to improve electrophoretic droplet sorting of single cells from doublets [101]. However, most uses of electrophoresis are in the detection of proteins, antibodies, and other cellular constituents.

4.6.3 Dielectrophoresis

Dielectrophoresis (DEP) is different from electrophoresis in that cells move in a non-uniform field as a result of their polarizability, not their charge. Non-uniform AC fields induce dipole moments across cells, effectively polarizing them and inducing migration. If the cells have a higher permeability than the fluid they are immersed in, they are attracted to the area with the highest field intensity, known as the maxima, in a process called positive DEP (pDEP). If the reverse is true and the cells have a lower permeability than the surrounding fluid, negative DEP (nDEP) will occur, and the cells will migrate away from the region of highest field intensity. The forces of dielectrophoresis depend on cellular properties such as size and shape, as well as the characteristics of the surrounding fluid and electric field.

DEP can be used to create cell traps via the application of a field gradient across planar electrodes. DEP traps are effectively strong enough to maintain cells in a desirable position despite external forces, such as fluid movement. Because different cell types have different dielectric properties, DEP can be used in the isolation of particular cell types of interest from heterogeneous cellular populations. DEP can also be used to separate cells and aggregates of different sizes within homogenous cellular populations [102–105].

While DEP has also been applied to numerous applications concerning blood, DNA, microorganisms, and other cells, this concept has not been as extensively investigated in cellular suspensions from dissociated tissues [106, 107]. However, initial progress has been made in the creation of a DEP microfluidic device using direct current to trap individual cells from cellular mixtures containing MCF7 and MDA-MB-231 cancer cells, as well as PBMC blood cells [108]. DEP devices can be structured so as to control the DEP at individual locations in a microfluidic chip, such as within specific wells. Lateral DEP devices can be used to position cells using interdigitated electrode arrays [109]. Feedback on cellular capture and location can be assessed by measuring changes in impedance as cells pass through different chip locations.

In addition to pDEP and nDEP, different subtypes of DEP exist, such as closed or conventional dielectrophoresis (cDEP) and traveling wave dielectrophoresis (twDEP) [110]. DEP is also widely multiplexed with droplet microfluidics, optical readout systems, and more, making it a very diverse and versatile technique.

4.6.4 Other electrical phenomena

Other electrical phenomena exist that can be helpful tools in single-cell manipulation and sorting, which could be potentially usefully integrated into single-cell processing workflows from tissue starting samples. The two most important examples of such phenomena are electroorientation and electrorotation.

In electroorientation, cells orient themselves so that they are aligned in a certain confirmation along with an applied electric field. This phenomenon occurs if the induced dipole moment along a particular axis of the cell is stronger, as the dipole moment has varying frequency dependency depending on the axis [17]. This effect is generally seen at higher frequencies, on the order of MHz, and is most frequently seen in non-human cells, although the principle has been applied in DEP systems to human cells as well [111].

In electrorotation, cells are able to be rotated by surrounding electrodes. The principle was first developed in Arnold and Zimmermann in 1988, in order to perform dielectric measurements on individual cells [112]. In this process, surrounding electrodes can be placed at the 0°, 90°, 180°, and 270° positions. The electrodes can emit constant electric fields, but generally use rotating electric fields that are phase-shifted by 90 degrees [17, 112]. As with electroorientation, this is a frequency dependent process.

4.7 Label-free sorting by acoustophoretic properties

Acoustophoresis is a process in which acoustic pressure waves are able to result in movement via standing or traveling wave formation. Acoustic techniques have been combined with microfluidic systems in recent years, creating a new subcategory of "acoustofluidic" devices. Acoustofluidic systems have been used extensively in cell sorting applications and offer benefits such as rapid movement over nm to mm length scales. Different acoustic waves that are used for this purpose consist of bulk standing waves, and surface acoustic waves (SAWs). SAWs are further subdivided into standing SAWs (SSAWs) and traveling SAWs (TSAWs). The key difference between bulk waves and standing waves are that bulk waves, generated by a piezoelectric transducer, propagate through the bulk of the microchannel chamber itself, while surface acoustic waves, generated by an interdigitated transducer, propagate through the bottom surface of the channel [113]. Acoustofluidic systems are frequently used to sort cells via deformability or other elastic properties.

4.7.1 Bulk standing waves

The first instance of bulk standing wave sorting was characterized by Johansson et al. in 2009, in which an ultrasonic transducer was used to facilitate microfluidic FACS [114]. In the bulk standing wave acoustophoretic process, pressure waves that are identical in terms of oscillation frequency and magnitude are applied to a fluid media in opposite directions, resulting in standing wave formation within the media. Standing waves contain nodes and oscillating antinodes, which represent 0% pressure fluctuation and 100% pressure fluctuation, respectively. In bulk standing waves, the transmitted acoustic wavelength matches the configuration and dimensions of a given channel, and particles are able to move outward. When placed in physiologically relevant buffers such as media, cells are able to be focused into areas within the microfluidic channel via this process as a result of their higher density than the surrounding fluid. They can be separated by size due to differences in forces that dominate at different particle sizes. Specifically, acoustic radiation dominates in larger particles while drag forces induced by acoustic streaming dominate in smaller ones [113].

The bulk acoustic wave separation technique has benefits over other acoustic techniques due to the increased flexibility in transducer placement, which makes microfluidic chip construction less complex. Bulk wave systems tend to operate using a comparatively lower frequency and longer wavelength, which gives them utility in the manipulation of particles of up to 200 μm, such as water-in-oil droplets [115]. Bulk wave mechanisms often require more energy than surface wave methods to facilitate cell sorting, due to the need to propagate signal throughout the entire bulk of the channel as opposed to the surface (**Figure 7a**) [116].

4.7.2 Surface acoustic waves

The study of SAWs dates back to 1885, when they were initially studied by Lord Rayleigh [117]. Surface acoustic waves travel parallel to the channel surface and are thought to be confined to within ~1 wavelength of the surface. SAWs are generated

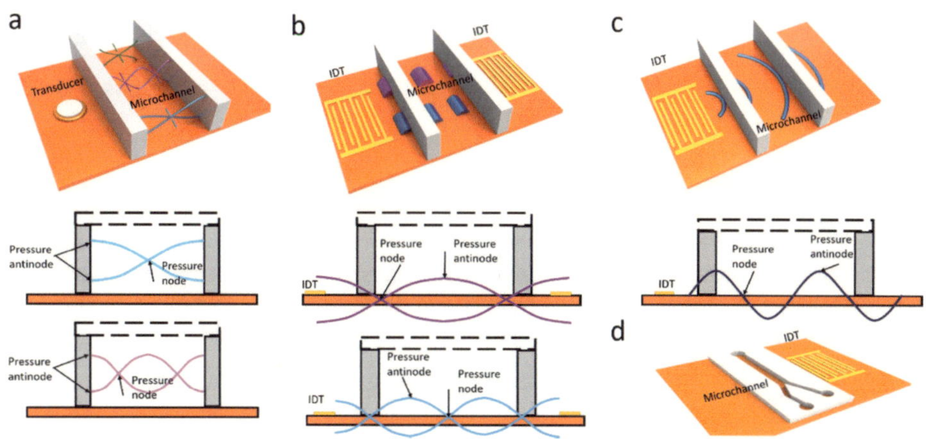

Figure 7.
Depiction of different acoustophoretic devices and concepts. (a) Depiction of bulk waves; (b) depiction of standing surface acoustic waves; (c) depiction of staveling surface acoustic waves; (d) illustration of an acoustic microfluidic device. IDT indicates interdigitated transistor. Reproduced with permission [113].

by a different transistor type - IDTs. SSAWs are characterized by fixed nodes and antinodes (**Figure 7b**). This occurs following the interference of two opposing traveling SAWs or a SAW that is reflected [113]. TSAWs, alternatively, are SAWs propagating in a single direction away from a single acoustic source (**Figure 7c**) [113]. Generally, SSAW approaches use two IDTs on opposite sides of a microfluidic device, while TSAW approaches use a single IDT single IDT (**Figure 7d**).

Particular cell types of interest can also be sorted in a contactless, label-free manner via deformation-assisted cell sorting using SAWs. SSAW fields can be applied across a channel by applying oscillation across interdigitated transducers at either side. Tunable IDT pitches give SAWs characteristic wavelengths, which can be determined using the simple equation (Eq. (1)). Where f is the frequency, v is the velocity, and x is the pitch of the transducer.

$$f = v / x \tag{1}$$

Surface acoustic wave platforms produce acoustic streaming when SAWs interact with surrounding fluid. In channels with SAW fields on either side (SSAW), hydrodynamic focusing occurs, which can also be referred to as "acoustic tweezers" [118]. The resulting streams move encapsulated cells or other particles, which can subsequently be trapped at pressure nodes along the chip [119]. For example, Nawaz et al. recently combined a real-time fluorescence and deformability cytometry method using SSAW-based label-free sorting [120]. This technique was found to reduce cell-processing time by an order of magnitude when compared to "state-of-the-art image-based sorters", without the need for specialized equipment.

Traveling surface acoustic waves (TSAWs) are less commonly used than standing surface acoustic waves (SSAWs), but have nevertheless been applied to facilitate cell sorting. Systems using traveling waves often employ them for the purpose of cellular deflection into collection wells or outlets via fluid streaming initiated by acoustic radiation forces [121, 122]. Cells can be sorted by size directly from bulk solution using a TSAWs, governed by a simple equation, adapted from the work of Skowronek et al., which describes whether acoustic radiation drag force from streaming dominates (Eq. (2)) [123]. In which K is a dimensionless factor, r is particle radius, and λ is the wavelength of the TSAW. When $K < 1$, streaming induced drag dominates. When $K > 1$, radiation dominates.

$$K = 2\pi r / \lambda \tag{2}$$

4.8 Other techniques

Certain other techniques can be used for cell sorting and manipulation that do not fit precisely into any of the aforementioned categories. An example that has garnered attention in recent years is bubble-based deflection, which can be produced by numerous different physical mechanisms.

4.8.1 Bubble-based deflection

Bubble deflection, also referred to by other terms such as 'vortex-actuated cell sorting' is a technique which uses instantaneous bubble-formation to in turn induce formation of a fluid jet which is then able to deflect cells of interest as they pass through a microchannel. Bubbles can be generated in one of several ways. The most common approach is to use heating to induce bubble formation. For example, resistive or electrical heating can be used to heat the surrounding area [124, 125]. Heating can also be created by optical mechanisms, such as laser light pulses [62, 125–128]. Ultrasound can also be used to create bubbles via cavitation, but, due to the large wavelengths, precise sorting of cells is difficult [129]. Spark discharge in dielectric liquids causes "electrical breakdown" and can induce the formation of highly pressurized bubbles at the electrodes [130].

Bubble deflection techniques are commonly integrated into a FACS workflow as a replacement step for electrostatic charging and charge-based deflection. In this standard workflow, optical detection with fluorophores is still needed. However, certain researchers have used bubbles as cellular tags, and measured displacement and velocity under ultrasound, representing a potentially label-free technique [131].

5. Conclusion

Sample preparation for single-cell analysis is a critical area of importance for future research in order to ensure widespread clinical translation of single-cell sequencing and other single-cell analysis techniques into cancer diagnostics and other workflows. The highlighted tissue dissociation, sample purification and cell manipulation techniques represent the current state-of-the art at the time of this writing for these processing steps.

New techniques relevant to the preparation of single cells from tissues will continue to be developed in the coming years. Additionally, researchers have begun combining several of the techniques discussed herein into new, hybrid devices, some of which have even been commercialized. For example, the CTC-iChip sorts cells using a combination of deterministic lateral displacement, inertial focusing, and magnetophoresis and was able to obtain a cell sorting rate of 10^7 cells/second and an enrichment yield of 97% rare cells [132]. While this device was notably tailored towards the liquid biopsy workflow, it's quite possible that integrated techniques targeting tissue dissociation into single cells will be forthcoming.

Comprehensive sample preparation instruments that can perform tissue dissociation, purification, and single-cell selection functions, among others are a current target of innovation, as evidenced by the work of Lombardo et al., who combined dissociation with purification via filters, enabling subsequent single-cell sequencing [15].

It is likely that machine learning and artificial intelligence will continue to be combined with the sample preparation process in order to improve the workflow. Laboratory automation techniques, such as automated liquid handling systems will also continue to increase in prominence to both heighten throughput and avoid manual processing. These methods will collectively expedite processing, making it significantly more efficient and less laborious.

The combination of high-resolution single-cell analysis techniques with high-heterogeneity tissue samples represents an ideal path forward to achieving next generation cancer diagnostics. However, for this future to become a reality, much progress and innovation is still needed within the tissue to single-cell sample preparation workflow. Developments are being made on multiple fronts, from dissociation and cellular manipulation to machine learning and computer vision. Integrated techniques are on the rise, and new devices are being commercialized. A comprehensive discussion of available technologies has been presented here with particular emphasis on translational potential within the tissue to the single-cell sequencing workflow.

6. Further reading

For more information on other steps in the sample preparation workflow, such as single-cell encapsulation and lysis, we recommend Hosic et al.'s comprehensive review [23]. To contextualize the relevance of the single-cell analysis workflow, the reader is also recommended the following papers on heterogeneity and spatial resolution in tissue as resolved through single-cell analysis [133–136]. For further information on experimental considerations for particular single-cell approaches, the reader is recommended the following resources: scDNAseq [137], scRNAseq [138–140], ATAC-seq [141], single-cell proteomics [142–144] and single-cell metabolomics [145, 146].

Acknowledgements

Figures 1-6 were created with BioRender.com.

Author details

E. Celeste Welch* and Anubhav Tripathi
Department of Biomedical Engineering, School of Engineering, Brown University, USA

*Address all correspondence to: cel_welch@brown.edu

References

[1] E. C. Welch, J. M. Powell, T. B. Clevinger, A. E. Fairman, and A. Shukla, "Advances in Biosensors and Diagnostic Technologies Using Nanostructures and Nanomaterials," *Adv. Funct. Mater.*, Jul. 2021, doi: 10.1002/adfm.202104126.

[2] A. Vlahiotis, B. Griffin, F. A. T. Stavros MD, and J. Margolis, "Analysis of utilization patterns and associated costs of the breast imaging and diagnostic procedures after screening mammography," *Clin. Outcomes Res.*, vol. Volume 10, Mar. 2018, doi: 10.2147/CEOR.S150260.

[3] Y. Wang and N. E. Navin, "Advances and Applications of Single-Cell Sequencing Technologies," *Mol. Cell*, vol. 58, no. 4, May 2015, doi: 10.1016/j.molcel.2015.05.005.

[4] G. POSTE, "Tissue dissociation with proteolytic enzymes *1Adsorption and activity of enzymes at the cell surface," *Exp. Cell Res.*, vol. 65, no. 2, Apr. 1971, doi: 10.1016/0014-4827(71)90014-0.

[5] J. D. Robin, W. E. Wright, Y. Zou, S. C. Cossette, M. W. Lawlor, and E. Gussoni, "Isolation and Immortalization of Patient-derived Cell Lines from Muscle Biopsy for Disease Modeling," *J. Vis. Exp.*, no. 95, Jan. 2015, doi: 10.3791/52307.

[6] E. C. Welch, H. Yu, and *A. Tripathi*, "Optimization of a Clinically Relevant Chemical-Mechanical Tissue Dissociation Workflow for Single-Cell Analysis," *Cell. Mol. Bioeng.*, Mar. 2021, doi: 10.1007/s12195-021-00667-y.

[7] S. Scheuermann, A. Schäfer, J. Langejürgen, and C. Reis, "A step towards enzyme-free tissue dissociation," *Curr. Dir. Biomed. Eng.*, vol. 5, no. 1, Sep. 2019, doi: 10.1515/cdbme-2019-0137.

[8] S. M. Hattersley, C. E. Dyer, J. Greenman, and S. J. Haswell, "Development of a microfluidic device for the maintenance and interrogation of viable tissue biopsies," *Lab Chip*, vol. 8, no. 11, 2008, doi: 10.1039/b809345h.

[9] C.-H. Lin, D.-C. Lee, H.-C. Chang, I.-M. Chiu, and C.-H. Hsu, "Single-Cell Enzyme-Free Dissociation of Neurospheres Using a Microfluidic Chip," *Anal. Chem.*, vol. 85, no. 24, Dec. 2013, doi: 10.1021/ac402724b.

[10] L. Wallman *et al.*, "Biogrid—a microfluidic device for large-scale enzyme-free dissociation of stem cell aggregates," *Lab Chip*, vol. 11, no. 19, 2011, doi: 10.1039/c1lc20316a.

[11] X. Qiu *et al.*, "Microfluidic filter device with nylon mesh membranes efficiently dissociates cell aggregates and digested tissue into single cells," *Lab Chip*, vol. 18, no. 18, 2018, doi: 10.1039/C8LC00507A.

[12] X. Qiu *et al.*, "Microfluidic channel optimization to improve hydrodynamic dissociation of cell aggregates and tissue," *Sci. Rep.*, vol. 8, no. 1, Dec. 2018, doi: 10.1038/s41598-018-20931-y.

[13] X. Qiu, J. De Jesus, M. Pennell, M. Troiani, and J. B. Haun, "Microfluidic device for mechanical dissociation of cancer cell aggregates into single cells," *Lab Chip*, vol. 15, no. 1, 2015, doi: 10.1039/C4LC01126K.

[14] X. Qiu *et al.*, "Microfluidic device for rapid digestion of tissues into cellular suspensions," *Lab Chip*, vol. 17, no. 19, 2017, doi: 10.1039/C7LC00575J.

[15] J. A. Lombardo, M. Aliaghaei, Q. H. Nguyen, K. Kessenbrock, and J. B. Haun, "Microfluidic platform accelerates tissue processing into single cells for molecular analysis and primary culture models," *Nat. Commun.*, vol. 12,

no. 1, Dec. 2021, doi: 10.1038/s41467-021-23238-1.

[16] E. C. Welch and A. Tripathi, "ELECTRICAL DISSOCIATION OF TISSUE SAMPLES INTO SINGLE CELLS AND/OR SMALLER GROUPS OF CELLS," 63/177,211, 2021.

[17] G. H. Markx, "The use of electric fields in tissue engineering," *Organogenesis*, vol. 4, no. 1, Jan. 2008, doi: 10.4161/org.5799.

[18] F. X. Hart and J. R. Palisano, "The Application of Electric Fields in Biology and Medicine," in *Electric Field*, InTech, 2018.

[19] J. P. Freyer and R. M. Sutherland, "Selective Dissociation and Characterization of Cells from Different Regions of Multicell Tumor Spheroids," *Cancer Res.*, vol. 40, no. 11, pp. 3956-3965, 1980.

[20] J. P. Freyer and P. L. Schor, "Automated Selective Dissociation of Cells From Different Regions of Multicellular Spheroids," *Vitr. Cell. Dev. Biol.*, vol. 25, no. 1, pp. 9-19, 1988.

[21] M. R. Emmert-Buck *et al.*, "Laser Capture Microdissection," *Science (80-.).*, vol. 274, no. 5289, Nov. 1996, doi: 10.1126/science.274.5289.998.

[22] R. F. Bonner, "CELL SAMPLING: Laser Capture Microdissection: Molecular Analysis of Tissue," *Science (80-.).*, vol. 278, no. 5342, Nov. 1997, doi: 10.1126/science.278.5342.1481.

[23] S. Hosic, S. K. Murthy, and A. N. Koppes, "Microfluidic Sample Preparation for Single Cell Analysis," *Anal. Chem.*, vol. 88, no. 1, Jan. 2016, doi: 10.1021/acs.analchem.5b04077.

[24] L. Moulédous, S. Hunt, R. Harcourt, J. Harry, K. L. Williams, and H. B. Gutstein, "Navigated laser capture microdissection as an alternative to direct histological staining for proteomic analysis of brain samples," *Proteomics*, vol. 3, no. 5, May 2003, doi: 10.1002/pmic.200300398.

[25] B. J. Erickson, P. Korfiatis, Z. Akkus, and T. L. Kline, "Machine Learning for Medical Imaging," *RadioGraphics*, vol. 37, no. 2, Mar. 2017, doi: 10.1148/rg.2017160130.

[26] B. Karaçali and A. Tözeren, "Automated detection of regions of interest for tissue microarray experiments: an image texture analysis," *BMC Med. Imaging*, vol. 7, no. 1, Dec. 2007, doi: 10.1186/1471-2342-7-2.

[27] J. Hipp *et al.*, "Computer-aided laser dissection: A microdissection workflow leveraging image analysis tools," *J. Pathol. Inform.*, vol. 9, no. 1, 2018, doi: 10.4103/jpi.jpi_60_18.

[28] S. Ren, Y. Luo, T. Yan, L. Wang, D. Chen, and X. Chen, "Machine learning-based automatic segmentation of region of interest in dynamic optical imaging," *AIP Adv.*, vol. 11, no. 1, Jan. 2021, doi: 10.1063/5.0033605.

[29] B. Maiorella, G. Dorin, A. Carion, and D. Harano, "Crossflow microfiltration of animal cells," *Biotechnol. Bioeng.*, vol. 37, no. 2, Jan. 1991, doi: 10.1002/bit.260370205.

[30] P. Preira, V. Grandné, J.-M. Forel, S. Gabriele, M. Camara, and O. Theodoly, "Passive circulating cell sorting by deformability using a microfluidic gradual filter," *Lab Chip*, vol. 13, no. 1, 2013, doi: 10.1039/C2LC40847C.

[31] L. A. Herzenberg, R. G. Sweet, and L. A. Herzenberg, "Fluorescence-activated cell sorting," *Sci. Am.*, vol. 234, no. 3, pp. 108-118, 1976.

[32] L. A. Herzenberg, D. Parks, B. Sahaf, O. Perez, M. Roederer, and L. A. Herzenberg, "The History and Future of the Fluorescence Activated Cell Sorter

and Flow Cytometry: A View from Stanford," *Clin. Chem.*, vol. 48, no. 10, Oct. 2002, doi: 10.1093/clinchem/48.10.1819.

[33] J. Krüger, K. Singh, A. O'Neill, C. Jackson, A. Morrison, and P. O'Brien, "Development of a microfluidic device for fluorescence activated cell sorting," *J. Micromechanics Microengineering*, vol. 12, no. 4, Jul. 2002, doi: 10.1088/0960-1317/12/4/324.

[34] B. Yao, G. Luo, X. Feng, W. Wang, L. Chen, and Y. Wang, "A microfluidic device based on gravity and electric force driving for flow cytometry and fluorescence activated cell sorting," *Lab Chip*, vol. 4, no. 6, 2004, doi: 10.1039/b408422e.

[35] J. Autebert *et al.*, "Microfluidic: An innovative tool for efficient cell sorting," *Methods*, vol. 57, no. 3, Jul. 2012, doi: 10.1016/j.ymeth.2012.07.002.

[36] J.-C. Baret *et al.*, "Fluorescence-activated droplet sorting (FADS): efficient microfluidic cell sorting based on enzymatic activity," *Lab Chip*, vol. 9, no. 13, 2009, doi: 10.1039/b902504a.

[37] H. M. Shapiro, *Practical flow cytometry*. John Wiley & Sons, 2005.

[38] O. Caen *et al.*, "High-throughput multiplexed fluorescence-activated droplet sorting," *Microsystems Nanoeng.*, vol. 4, no. 1, Dec. 2018, doi: 10.1038/s41378-018-0033-2.

[39] I. Pirozzi, A. Snider, M. Kraus, E. R. Schönbrunner, and *A. Tripathi*, "Microfluidic Immiscible Phase Filtration System for the Isolation of Small Numbers of Cells from Whole Blood," *Cytom. Part A*, vol. 95, no. 8, Aug. 2019, doi: 10.1002/cyto.a.23736.

[40] S. Miltenyi, W. Müller, W. Weichel, and A. Radbruch, "High gradient magnetic cell separation with MACS," *Cytometry*, vol. 11, no. 2, 1990, doi: 10.1002/cyto.990110203.

[41] V. Martin, Siewert C, Scharl A, and Harms T, "Immunomagnetic enrichment of disseminated epithelial tumor cells from peripheral blood by MACS."

[42] A. Palmon, R. David, Y. Neumann, R. Stiubea-Cohen, G. Krief, and D. J. Aframian, "High-efficiency immunomagnetic isolation of solid tissue-originated integrin-expressing adult stem cells," *Methods*, vol. 56, no. 2, Feb. 2012, doi: 10.1016/j.ymeth.2011.10.002.

[43] N. HARBECK *et al.*, "MODEL SYSTEM FOR ISOLATION OF COMPETENT OVARIAN-CARCINOMA CELLS FROM FRESH TUMOR-TISSUE BY A MAGNETIC SEPARATION TECHNIQUE (MACS)," *Int. J. Oncol.*, Jun. 1995, doi: 10.3892/ijo.6.6.1249.

[44] Y. Wang *et al.*, "A microfluidic robot for rare cell sorting based on machine vision identification and multi-step sorting strategy," *Talanta*, vol. 226, May 2021, doi: 10.1016/j.talanta.2021.122136.

[45] A. Ashkin, J. M. Dziedzic, J. E. Bjorkholm, and S. Chu, "Observation of a single-beam gradient force optical trap for dielectric particles," *Opt. Lett.*, vol. 11, no. 5, May 1986, doi: 10.1364/OL.11.000288.

[46] X. Wang *et al.*, "Enhanced cell sorting and manipulation with combined optical tweezer and microfluidic chip technologies," *Lab Chip*, vol. 11, no. 21, 2011, doi: 10.1039/c1lc20653b.

[47] Y. Z. Yoon and P. Cicuta, "Optical trapping of colloidal particles and cells by focused evanescent fields using conical lenses," *Opt. Express*, vol. 18, no. 7, Mar. 2010, doi: 10.1364/OE.18.007076.

[48] Y. Li, Z. Guo, and S. Qu, "Living cell manipulation in a microfluidic device by femtosecond optical tweezers," *Opt. Lasers Eng.*, vol. 55, Apr. 2014, doi: 10.1016/j.optlaseng.2013.11.001.

[49] T. Yamazaki *et al.*, "Construction of 3D Cellular Composites with Stem Cells Derived from Adipose Tissue and Endothelial Cells by Use of Optical Tweezers in a Natural Polymer Solution," *Materials (Basel).*, vol. 12, no. 11, May 2019, doi: 10.3390/ma12111759.

[50] J. P. K. Armstrong and M. M. Stevens, "Using Remote Fields for Complex Tissue Engineering," *Trends Biotechnol.*, vol. 38, no. 3, Mar. 2020, doi: 10.1016/j.tibtech.2019.07.005.

[51] X. Li, C. Liu, S. Chen, Y. Wang, S. H. Cheng, and D. Sun, "<italic>In Vivo</italic> Manipulation of Single Biological Cells With an Optical Tweezers-Based Manipulator and a Disturbance Compensation Controller," *IEEE Trans. Robot.*, vol. 33, no. 5, Oct. 2017, doi: 10.1109/TRO.2017.2718554.

[52] X. Li, H. Yang, H. Huang, and D. Sun, "A switching controller for high speed cell transportation by using a robot-aided optical tweezers system," *Automatica*, vol. 89, Mar. 2018, doi: 10.1016/j.automatica.2017.11.014.

[53] M. Xie, A. Shakoor, and C. Wu, "Manipulation of Biological Cells Using a Robot-Aided Optical Tweezers System," *Micromachines*, vol. 9, no. 5, May 2018, doi: 10.3390/mi9050245.

[54] M. C. Wu, "Optoelectronic tweezers," *Nat. Photonics*, vol. 5, no. 6, Jun. 2011, doi: 10.1038/nphoton.2011.98.

[55] A. T. Ohta *et al.*, "Dynamic Cell and Microparticle Control via Optoelectronic Tweezers," *J. Microelectromechanical Syst.*, vol. 16, no. 3, Jun. 2007, doi: 10.1109/JMEMS.2007.896717.

[56] P. Y. Chiou, A. T. Ohta, and M. C. Wu, "Massively parallel manipulation of single cells and microparticles using optical images," *Nature*, vol. 436, no. 7049, Jul. 2005, doi: 10.1038/nature03831.

[57] A. T. Ohta *et al.*, "Optically Controlled Cell Discrimination and Trapping Using Optoelectronic Tweezers," *IEEE J. Sel. Top. Quantum Electron.*, vol. 13, no. 2, 2007, doi: 10.1109/JSTQE.2007.893558.

[58] L.-Y. Ke *et al.*, "Cancer immunotherapy μ-environment LabChip: taking advantage of optoelectronic tweezers," *Lab Chip*, vol. 18, no. 1, 2018, doi: 10.1039/C7LC00963A.

[59] S. Zhang *et al.*, "The optoelectronic microrobot: A versatile toolbox for micromanipulation," *Proc. Natl. Acad. Sci.*, vol. 116, no. 30, Jul. 2019, doi: 10.1073/pnas.1903406116.

[60] K. Ramser and D. Hanstorp, "Optical manipulation for single-cell studies," *J. Biophotonics*, vol. 3, no. 4, Apr. 2010, doi: 10.1002/jbio.200910050.

[61] S.-K. Hoi, C. Udalagama, C.-H. Sow, F. Watt, and A. A. Bettiol, "Microfluidic sorting system based on optical force switching," *Appl. Phys. B*, vol. 97, no. 4, Dec. 2009, doi: 10.1007/s00340-009-3687-5.

[62] T.-H. Wu *et al.*, "Pulsed laser triggered high speed microfluidic fluorescence activated cell sorter," *Lab Chip*, vol. 12, no. 7, 2012, doi: 10.1039/c2lc21084c.

[63] C. M. Bailey-Hytholt, S. Sayeed, M. Kraus, R. Joseph, A. Shukla, and *A. Tripathi*, "A Rapid Method for Label-Free Enrichment of Rare Trophoblast Cells from Cervical Samples," *Sci. Rep.*, vol. 9, no. 1, Dec. 2019, doi: 10.1038/s41598-019-48346-3.

[64] C. M. Bailey-Hytholt, S. Sayeed, A. Shukla, and *A. Tripathi*, "Enrichment of Placental Trophoblast Cells from Clinical Cervical Samples Using Differences in Surface Adhesion on an Inclined Plane," *Ann. Biomed. Eng.*, Mar. 2021, doi: 10.1007/s10439-021-02742-x.

[65] J. D. Amack and M. L. Manning, "Knowing the Boundaries: Extending the Differential Adhesion Hypothesis in Embryonic Cell Sorting," *Science (80-.).*, vol. 338, no. 6104, Oct. 2012, doi: 10.1126/science.1223953.

[66] S. L. Stott *et al.*, "Isolation of circulating tumor cells using a microvortex-generating herringbone-chip," *Proc. Natl. Acad. Sci.*, vol. 107, no. 43, Oct. 2010, doi: 10.1073/pnas.1012539107.

[67] Y. Chikaishi, K. Yoneda, T. Ohnaga, and F. Tanaka, "EpCAM-independent capture of circulating tumor cells with a 'universal CTC-chip,'" *Oncol. Rep.*, vol. 37, no. 1, Jan. 2017, doi: 10.3892/or.2016.5235.

[68] K. Pu, C. Li, N. Zhang, H. Wang, W. Shen, and Y. Zhu, "Epithelial cell adhesion molecule independent capture of non-small lung carcinoma cells with peptide modified microfluidic chip," *Biosens. Bioelectron.*, vol. 89, Mar. 2017, doi: 10.1016/j.bios.2016.09.092.

[69] L. R. Huang, "Continuous Particle Separation Through Deterministic Lateral Displacement," *Science (80-.).*, vol. 304, no. 5673, May 2004, doi: 10.1126/science.1094567.

[70] J. McGrath, M. Jimenez, and H. Bridle, "Deterministic lateral displacement for particle separation: a review," *Lab Chip*, vol. 14, no. 21, 2014, doi: 10.1039/C4LC00939H.

[71] A. Hochstetter *et al.*, "Deterministic Lateral Displacement: Challenges and Perspectives," *ACS Nano*, vol. 14, no. 9, Sep. 2020, doi: 10.1021/acsnano.0c05186.

[72] J. V. Green, M. Radisic, and S. K. Murthy, "Deterministic Lateral Displacement as a Means to Enrich Large Cells for Tissue Engineering," *Anal. Chem.*, vol. 81, no. 21, Nov. 2009, doi: 10.1021/ac9018395.

[73] Z. Liu *et al.*, "Rapid isolation of cancer cells using microfluidic deterministic lateral displacement structure," *Biomicrofluidics*, vol. 7, no. 1, Jan. 2013, doi: 10.1063/1.4774308.

[74] K. Loutherback, K. S. Chou, J. Newman, J. Puchalla, R. H. Austin, and J. C. Sturm, "Improved performance of deterministic lateral displacement arrays with triangular posts," *Microfluid. Nanofluidics*, vol. 9, no. 6, Dec. 2010, doi: 10.1007/s10404-010-0635-y.

[75] S. Choi, J. M. Karp, and R. Karnik, "Cell sorting by deterministic cell rolling," *Lab Chip*, vol. 12, no. 8, 2012, doi: 10.1039/c2lc21225k.

[76] D. Di Carlo, D. Irimia, R. G. Tompkins, and M. Toner, "Continuous inertial focusing, ordering, and separation of particles in microchannels," *Proc. Natl. Acad. Sci.*, vol. 104, no. 48, Nov. 2007, doi: 10.1073/pnas.0704958104.

[77] J. M. Martel and M. Toner, "Inertial focusing dynamics in spiral microchannels," *Phys. Fluids*, vol. 24, no. 3, Mar. 2012, doi: 10.1063/1.3681228.

[78] A. Al-Halhouli *et al.*, "Enhanced inertial focusing of microparticles and cells by integrating trapezoidal microchambers in spiral microfluidic channels," *RSC Adv.*, vol. 9, no. 33, 2019, doi: 10.1039/C9RA03587G.

[79] A. Shiriny and M. Bayareh, "Inertial focusing of CTCs in a novel spiral microchannel," *Chem. Eng. Sci.*, vol. 229, Jan. 2021, doi: 10.1016/j.ces.2020.116102.

[80] Y. Gou *et al.*, "Sheathless Inertial Focusing Chip Combining a Spiral Channel with Periodic Expansion Structures for Efficient and Stable Particle Sorting," *Anal. Chem.*, vol. 92, no. 2, Jan. 2020, doi: 10.1021/acs.analchem.9b03692.

[81] H. Ren *et al.*, "Multiplexed serpentine microchannels for

high-throughput sorting of disseminated tumor cells from malignant pleural effusion," *Sensors Actuators B Chem.*, vol. 337, Jun. 2021, doi: 10.1016/j.snb.2021.129758.

[82] D. Tang, M. Chen, Y. Han, N. Xiang, and Z. Ni, "Asymmetric serpentine microchannel based impedance cytometer enabling consistent transit and accurate characterization of tumor cells and blood cells," *Sensors Actuators B Chem.*, vol. 336, Jun. 2021, doi: 10.1016/j.snb.2021.129719.

[83] V. Choesmel *et al.*, "Enrichment methods to detect bone marrow micrometastases in breast carcinoma patients: clinical relevance," *Breast Cancer Res.*, vol. 6, no. 5, Oct. 2004, doi: 10.1186/bcr898.

[84] R. Gorkin *et al.*, "Centrifugal microfluidics for biomedical applications," *Lab Chip*, vol. 10, no. 14, 2010, doi: 10.1039/b924109d.

[85] A. Snider, I. Pirozzi, and *A. Tripathi*, "Centrifugal Microfluidics Traps for Parallel Isolation and Imaging of Single Cells," *Micromachines*, vol. 11, no. 2, Jan. 2020, doi: 10.3390/mi11020149.

[86] O. Strohmeier *et al.*, "Centrifugal microfluidic platforms: advanced unit operations and applications," *Chem. Soc. Rev.*, vol. 44, no. 17, 2015, doi: 10.1039/C4CS00371C.

[87] X. Zhang, Y. Li, S. Shen, S. Lee, and H. Dou, "Field-flow fractionation: A gentle separation and characterization technique in biomedicine," *TrAC Trends Anal. Chem.*, vol. 108, Nov. 2018, doi: 10.1016/j.trac.2018.09.005.

[88] C. Mélin *et al.*, "Cancer Stem Cell Sorting from Colorectal Cancer Cell Lines by Sedimentation Field Flow Fractionation," *Anal. Chem.*, vol. 84, no. 3, Feb. 2012, doi: 10.1021/ac202797z.

[89] S. Wang *et al.*, "Microalgae separation by inertia-enhanced pinched flow fractionation," *Electrophoresis*, May 2021, doi: 10.1002/elps.202000325.

[90] J. T. W. Berendsen, J. C. T. Eijkel, A. M. Wetzels, and L. I. Segerink, "Separation of spermatozoa from erythrocytes using their tumbling mechanism in a pinch flow fractionation device," *Microsystems Nanoeng.*, vol. 5, no. 1, Dec. 2019, doi: 10.1038/s41378-019-0068-z.

[91] X.-B. Wang, J. Yang, Y. Huang, J. Vykoukal, F. F. Becker, and P. R. C. Gascoyne, "Cell Separation by Dielectrophoretic Field-flow-fractionation," *Anal. Chem.*, vol. 72, no. 4, Feb. 2000, doi: 10.1021/ac990922o.

[92] A. Shamloo and A. Kamali, "Numerical analysis of a dielectrophoresis field-flow fractionation device for the separation of multiple cell types," *J. Sep. Sci.*, vol. 40, no. 20, Oct. 2017, doi: 10.1002/jssc.201700325.

[93] C. F. Gonzalez and V. T. Remcho, "Harnessing dielectric forces for separations of cells, fine particles and macromolecules," *J. Chromatogr. A*, vol. 1079, no. 1-2, Jun. 2005, doi: 10.1016/j.chroma.2005.03.070.

[94] C. Wyatt Shields IV, C. D. Reyes, and G. P. López, "Microfluidic cell sorting: a review of the advances in the separation of cells from debulking to rare cell isolation," *Lab Chip*, vol. 15, no. 5, 2015, doi: 10.1039/C4LC01246A.

[95] B. Yan, B. Chen, F. Liu, J. Wu, and Y. Xiong, "Combining field-modulating electroosmotic vortex and insulating post to manipulate particles based on dielectrophoresis," *Appl. Math. Mech.*, vol. 42, no. 3, Mar. 2021, doi: 10.1007/s10483-021-2706-5.

[96] L. Li, X. Wang, Q. Pu, and S. Liu, "Advancement of electroosmotic pump

in microflow analysis: A review," *Anal. Chim. Acta*, vol. 1060, Jul. 2019, doi: 10.1016/j.aca.2019.02.004.

[97] M. Alipanah, M. Hafttananian, N. Hedayati, A. Ramiar, and M. Alipanah, "Microfluidic on-demand particle separation using induced charged electroosmotic flow and magnetic field," *J. Magn. Magn. Mater.*, vol. 537, Nov. 2021, doi: 10.1016/j.jmmm.2021.168156.

[98] S. Valagerahally Puttaswamy, S. Sivashankar, C.-H. Yeh, R.-J. Chen, and C. H. Liu, "Electrodynamically actuated on-chip flow cytometry with low shear stress for electro-osmosis based sorting using low conductive medium," *Microelectron. Eng.*, vol. 87, no. 12, Dec. 2010, doi: 10.1016/j.mee.2010.07.023.

[99] X. Ou, *P. Chen*, X. Huang, S. Li, and B. Liu, "Microfluidic chip electrophoresis for biochemical analysis," *J. Sep. Sci.*, vol. 43, no. 1, Jan. 2020, doi: 10.1002/jssc.201900758.

[100] W. Zhang, N. Li, H. Zeng, H. Nakajima, J.-M. Lin, and K. Uchiyama, "Inkjet Printing Based Separation of Mammalian Cells by Capillary Electrophoresis," *Anal. Chem.*, vol. 89, no. 17, Sep. 2017, doi: 10.1021/acs.analchem.7b02624.

[101] F. Guo *et al.*, "Droplet electric separator microfluidic device for cell sorting," *Appl. Phys. Lett.*, vol. 96, no. 19, May 2010, doi: 10.1063/1.3360812.

[102] J. Yoshioka, Y. Ohsugi, T. Yoshitomi, T. Yasukawa, N. Sasaki, and K. Yoshimoto, "Label-Free Rapid Separation and Enrichment of Bone Marrow-Derived Mesenchymal Stem Cells from a Heterogeneous Cell Mixture Using a Dielectrophoresis Device," *Sensors*, vol. 18, no. 9, Sep. 2018, doi: 10.3390/s18093007.

[103] M. Elitas, Y. Yildizhan, M. Islam, R. Martinez-Duarte, and D. Ozkazanc, "Dielectrophoretic characterization and separation of monocytes and macrophages using 3D carbon-electrodes," *Electrophoresis*, vol. 40, no. 2, Jan. 2019, doi: 10.1002/elps.201800324.

[104] K. Kikkeri, B. A. Kerr, A. S. Bertke, J. S. Strobl, and M. Agah, "Passivated-electrode insulator-based dielectrophoretic separation of heterogeneous cell mixtures," *J. Sep. Sci.*, vol. 43, no. 8, Apr. 2020, doi: 10.1002/jssc.201900553.

[105] Y.-C. Kung, K. R. Niazi, and P.-Y. Chiou, "Tunnel dielectrophoresis for ultra-high precision size-based cell separation," *Lab Chip*, vol. 21, no. 6, 2021, doi: 10.1039/D0LC00853B.

[106] A. Sonnenberg *et al.*, "Dielectrophoretic isolation and detection of cfc-DNA nanoparticulate biomarkers and virus from blood," *Electrophoresis*, vol. 34, no. 7, Apr. 2013, doi: 10.1002/elps.201200444.

[107] F. S. O. Fritzsch *et al.*, "Picoliter nDEP traps enable time-resolved contactless single bacterial cell analysis in controlled microenvironments," *Lab Chip*, vol. 13, no. 3, 2013, doi: 10.1039/C2LC41092C.

[108] S. Bhattacharya *et al.*, "Selective trapping of single mammalian breast cancer cells by insulator-based dielectrophoresis," *Anal. Bioanal. Chem.*, vol. 406, no. 7, Mar. 2014, doi: 10.1007/s00216-013-7598-2.

[109] L. Wang, L. A. Flanagan, N. L. Jeon, E. Monuki, and A. P. Lee, "Dielectrophoresis switching with vertical sidewall electrodes for microfluidic flow cytometry," *Lab Chip*, vol. 7, no. 9, 2007, doi: 10.1039/b705386j.

[110] Y. Wu, Y. Ren, Y. Tao, and H. Jiang, "Fluid pumping and cells separation by DC-biased traveling wave electroosmosis and dielectrophoresis," *Microfluid. Nanofluidics*, vol. 21, no. 3,

Mar. 2017, doi: 10.1007/s10404-017-1862-2.

[111] M. Yang and X. Zhang, "Electrical assisted patterning of cardiac myocytes with controlled macroscopic anisotropy using a microfluidic dielectrophoresis chip," *Sensors Actuators A Phys.*, vol. 135, no. 1, Mar. 2007, doi: 10.1016/j.sna.2006.06.071.

[112] W. M. Arnold and U. Zimmermann, "Electro-rotation: development of a technique for dielectric measurements on individual cells and particles," *J. Electrostat.*, vol. 21, no. 2-3, Sep. 1988, doi: 10.1016/0304-3886(88)90027-7.

[113] Y. Gao, M. Wu, Y. Lin, and J. Xu, "Acoustic Microfluidic Separation Techniques and Bioapplications: A Review," *Micromachines*, vol. 11, no. 10, Oct. 2020, doi: 10.3390/mi11100921.

[114] L. Johansson, F. Nikolajeff, S. Johansson, and S. Thorslund, "On-Chip Fluorescence-Activated Cell Sorting by an Integrated Miniaturized Ultrasonic Transducer," *Anal. Chem.*, vol. 81, no. 13, Jul. 2009, doi: 10.1021/ac802681r.

[115] I. Leibacher, P. Reichert, and J. Dual, "Microfluidic droplet handling by bulk acoustic wave (BAW) acoustophoresis," *Lab Chip*, vol. 15, no. 13, 2015, doi: 10.1039/C5LC00083A.

[116] G. P. Gautam, T. Burger, A. Wilcox, M. J. Cumbo, S. W. Graves, and M. E. Piyasena, "Simple and inexpensive micromachined aluminum microfluidic devices for acoustic focusing of particles and cells," *Anal. Bioanal. Chem.*, vol. 410, no. 14, May 2018, doi: 10.1007/s00216-018-1034-6.

[117] Lord Rayleigh, "On Waves Propagated along the Plane Surface of an Elastic Solid," *Proc. London Math. Soc.*, vol. s1-17, no. 1, Nov. 1885, doi: 10.1112/plms/s1-17.1.4.

[118] X. Ding *et al.*, "On-chip manipulation of single microparticles, cells, and organisms using surface acoustic waves," *Proc. Natl. Acad. Sci.*, vol. 109, no. 28, Jul. 2012, doi: 10.1073/pnas.1209288109.

[119] S.-C. S. Lin, X. Mao, and T. J. Huang, "Surface acoustic wave (SAW) acoustophoresis: now and beyond," *Lab Chip*, vol. 12, no. 16, 2012, doi: 10.1039/c2lc90076a.

[120] A. A. Nawaz *et al.*, "Intelligent image-based deformation-assisted cell sorting with molecular specificity," *Nat. Methods*, vol. 17, no. 6, Jun. 2020, doi: 10.1038/s41592-020-0831-y.

[121] L. Schmid, D. A. Weitz, and T. Franke, "Sorting drops and cells with acoustics: acoustic microfluidic fluorescence-activated cell sorter," *Lab Chip*, vol. 14, no. 19, 2014, doi: 10.1039/C4LC00588K.

[122] G. Destgeer, B. H. Ha, J. Park, J. H. Jung, A. Alazzam, and H. J. Sung, "Travelling Surface Acoustic Waves Microfluidics," *Phys. Procedia*, vol. 70, 2015, doi: 10.1016/j.phpro.2015.08.028.

[123] V. Skowronek, R. W. Rambach, L. Schmid, K. Haase, and T. Franke, "Particle Deflection in a Poly(dimethylsiloxane) Microchannel Using a Propagating Surface Acoustic Wave: Size and Frequency Dependence," *Anal. Chem.*, vol. 85, no. 20, Oct. 2013, doi: 10.1021/ac402607p.

[124] K. de Wijs *et al.*, "Micro vapor bubble jet flow for safe and high-rate fluorescence-activated cell sorting," *Lab Chip*, vol. 17, no. 7, 2017, doi: 10.1039/C6LC01560C.

[125] R. H. Pritchard *et al.*, "Cell sorting actuated by a microfluidic inertial vortex," *Lab Chip*, vol. 19, no. 14, 2019, doi: 10.1039/C9LC00120D.

[126] B. Han, K. Köhler, K. Jungnickel, R. Mettin, W. Lauterborn, and A. Vogel,

"Dynamics of laser-induced bubble pairs," *J. Fluid Mech.*, vol. 771, May 2015, doi: 10.1017/jfm.2015.183.

[127] Y. Chen, T.-H. Wu, Y.-C. Kung, M. A. Teitell, and P.-Y. Chiou, "3D pulsed laser-triggered high-speed microfluidic fluorescence-activated cell sorter," *Analyst*, vol. 138, no. 24, 2013, doi: 10.1039/c3an01266b.

[128] Y. Chen, A. J. Chung, T.-H. Wu, M. A. Teitell, D. Di Carlo, and P.-Y. Chiou, "Pulsed Laser Activated Cell Sorting with Three Dimensional Sheathless Inertial Focusing," *Small*, vol. 10, no. 9, May 2014, doi: 10.1002/smll.201302885.

[129] C. Lee *et al.*, "Microfluidic droplet sorting with a high frequency ultrasound beam," *Lab Chip*, vol. 12, no. 15, 2012, doi: 10.1039/c2lc21123h.

[130] J. Zhao and Z. You, "Spark-generated microbubble cell sorter for microfluidic flow cytometry," *Cytom. Part A*, vol. 93, no. 2, Feb. 2018, doi: 10.1002/cyto.a.23296.

[131] T. J. Matula *et al.*, "Ultrasound-based cell sorting with microbubbles: A feasibility study," *J. Acoust. Soc. Am.*, vol. 144, no. 1, Jul. 2018, doi: 10.1121/1.5044405.

[132] N. M. Karabacak *et al.*, "Microfluidic, marker-free isolation of circulating tumor cells from blood samples," *Nat. Protoc.*, vol. 9, no. 3, Mar. 2014, doi: 10.1038/nprot.2014.044.

[133] Q. F. Wills *et al.*, "Single-cell gene expression analysis reveals genetic associations masked in whole-tissue experiments," *Nat. Biotechnol.*, vol. 31, no. 8, Aug. 2013, doi: 10.1038/nbt.2642.

[134] M. Schelker *et al.*, "Estimation of immune cell content in tumour tissue using single-cell RNA-seq data," *Nat. Commun.*, vol. 8, no. 1, Dec. 2017, doi: 10.1038/s41467-017-02289-3.

[135] X. Wang *et al.*, "Three-dimensional intact-tissue sequencing of single-cell transcriptional states," *Science (80-.).*, vol. 361, no. 6400, Jul. 2018, doi: 10.1126/science.aat5691.

[136] K. Achim *et al.*, "High-throughput spatial mapping of single-cell RNA-seq data to tissue of origin," *Nat. Biotechnol.*, vol. 33, no. 5, May 2015, doi: 10.1038/nbt.3209.

[137] H. Zahn *et al.*, "Scalable whole-genome single-cell library preparation without preamplification," *Nat. Methods*, vol. 14, no. 2, Feb. 2017, doi: 10.1038/nmeth.4140.

[138] Q. H. Nguyen, N. Pervolarakis, K. Nee, and K. Kessenbrock, "Experimental Considerations for Single-Cell RNA Sequencing Approaches," *Front. Cell Dev. Biol.*, vol. 6, Sep. 2018, doi: 10.3389/fcell.2018.00108.

[139] A. Lafzi, C. Moutinho, S. Picelli, and H. Heyn, "Tutorial: guidelines for the experimental design of single-cell RNA sequencing studies," *Nat. Protoc.*, vol. 13, no. 12, Dec. 2018, doi: 10.1038/s41596-018-0073-y.

[140] Y. Bykov, S. H. Kim, and D. Zamarin, "Preparation of single cells from tumors for single-cell RNA sequencing," 2020.

[141] M. R. Corces *et al.*, "An improved ATAC-seq protocol reduces background and enables interrogation of frozen tissues," *Nat. Methods*, vol. 14, no. 10, Oct. 2017, doi: 10.1038/nmeth.4396.

[142] W. Liu *et al.*, "Sample preparation method for isolation of single-cell types from mouse liver for proteomic studies," *Proteomics*, vol. 11, no. 17, Sep. 2011, doi: 10.1002/pmic.201100157.

[143] M. Dou *et al.*, "High-Throughput Single Cell Proteomics Enabled by Multiplex Isobaric Labeling in a

Nanodroplet Sample Preparation Platform," *Anal. Chem.*, vol. 91, no. 20, Oct. 2019, doi: 10.1021/acs.analchem.9b03349.

[144] S. M. Williams *et al.*, "Automated Coupling of Nanodroplet Sample Preparation with Liquid Chromatography–Mass Spectrometry for High-Throughput Single-Cell Proteomics," *Anal. Chem.*, vol. 92, no. 15, Aug. 2020, doi: 10.1021/acs.analchem.0c01551.

[145] K. D. Duncan, J. Fyrestam, and I. Lanekoff, "Advances in mass spectrometry based single-cell metabolomics," *Analyst*, vol. 144, no. 3, 2019, doi: 10.1039/C8AN01581C.

[146] A. Ali *et al.*, "Single-cell metabolomics by mass spectrometry: Advances, challenges, and future applications," *TrAC Trends Anal. Chem.*, vol. 120, Nov. 2019, doi: 10.1016/j.trac.2019.02.033.

Chapter 5

Particle and Cell Separation

J. Paul Robinson

Abstract

Many processors are available for separating particles and/or cells, but few can match the capacity of flow cytometry – in particular the sorting component. Several aspects unique to cell sorting give it such power. First, particles can be separated based on size, complexity, fluorescence, or any combination of these parameters. Second, it is entirely possible to separate particles under sterile conditions, making this technology very advantageous for selecting cells for culture. Third, when this sterile environment is combined with a highly controlled safety system, it is possible to safely sort and separate highly pathogenic organisms or even cells containing such pathogens. The very latest instruments available add even more power by introducing the ability to sort cells based on spectral unmixing. This last option requires incredible computer power and very-high-speed processing, since the sort decision is based on computational algorithms derived from the spectral mixture being analyzed.

Keywords: flow cytometry, sorting, single-cell analysis, fluorescence, light scatter, multiparameter analysis

1. Introduction

Available for over 50 years, cell sorting is now a technology undergoing significant change as new approaches emerge. The first effective cell sorter that demonstrated an ability to sort a significant number of particles or cells was developed by Mack Fulwyler in 1965 [1]. The foundation of the technology was the ultimate result of a publication by a pathologist named Lushbaugh that purported to identify multiple populations of red blood cells based on measurement of Coulter volume [2]. Fulwyler determined from an engineering perspective that Lushbaugh had incorrectly used the Coulter counter and set out to prove that red blood cells were not separate populations by designing an instrument that could physically separate individual populations. A few years later, Len Herzenberg expanded and developed the technology [3, 4] primarily for the immunology community as a fundamental tool in fluorescence analysis of cells [5–8].

Since those early days, sorting technology has been used in every aspect of science. Early practical applications of flow cytometry were for analysis of sperm [9], determination of fertility potential [10], and separating female and male sperm for breeding purposes [11]. Sorting was subsequently used for determination of DNA content [12], separation of human chromosomes [13], separation of live cells from dead ones [14], sorting of hemopoietic stem cells [15], and isolation of cloned isotype switch variants by fluorescence cell sorting [16]. All these were in the ten years following release of the first commercial instrument. Since that time, every conceivable particle and/or cell type has been physically sorted by flow, making it one of the most important scientific technologies in our armory.

2. The evolution of cell sorting

Cell sorting is the most efficient method currently available for creating single-cell populations. An advantage of flow cytometry is that this process can be achieved without damaging the cells to any significant extent, and in a sterile environment to boot.

2.1 Coulter volume–based sorting

The original device developed by Fulwyler was based on sorting by Coulter volume, as illustrated in **Figure 1**. The concept as originally defined was that cells of different volumes would be separable because the impedance varied with cell volume. This method of sorting provides only a single parameter, Coulter volume, and thus does not require a light source as later flow-cytometry technologies did. Interestingly, Fulwyler's techniques, while they created the stimulus for an entire industry, did not become integrated into the field of cell separation for almost 40 years, when microfluidic systems saw huge advantages in impedance-based separation (discussed below).

2.2 Principles of piezo-based cell sorting

The most common approach for current cell sorts is the use of piezo crystals to general a high-frequency vibration within the sorting chamber. Because the entire chamber vibrates, the stream emanating from the nozzle breaks up into droplets. These droplets contain the item of interest for sorting – of course it is also possible that the droplet fails to contain the particle or contains more than one. Separating a particular droplet from the rest of the stream results in the sorting process for any particle or cell that is within that droplet. **Figure 2** shows an overview of the process.

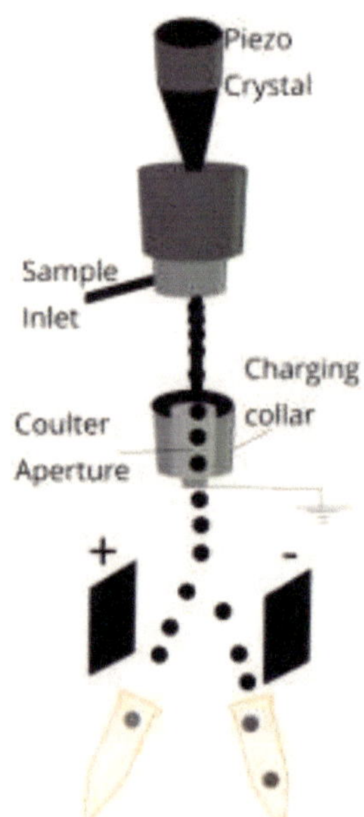

Figure 1.
Sketch of the original Fulwyler impedance-based cell sorter. This instrument had a single channel of data – Impedance. Reproduced with permission: J. Paul Robinson - Purdue University Cytometry Laboratories.

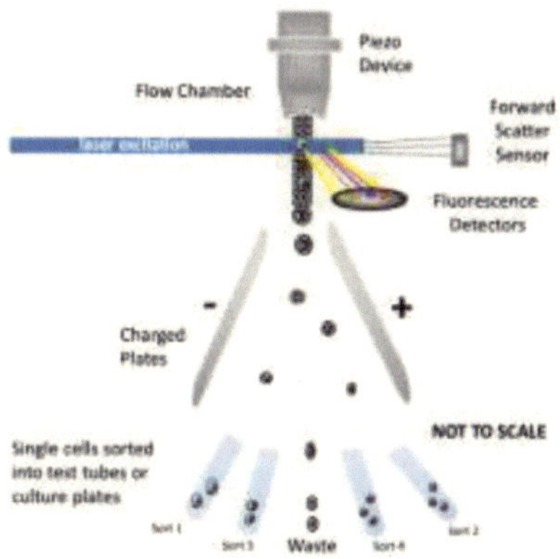

Figure 2.
Overview of the typical setup for a piezo-based cell or particle sorter. Shown at the top is a piezo device that causes the entire chamber to vibrate at high frequencies. This results in the stream's breaking into single droplets, each of which can be individually charged and then pulled toward a charged plate. Reproduced with permission: J. Paul Robinson - Purdue University Cytometry Laboratories.

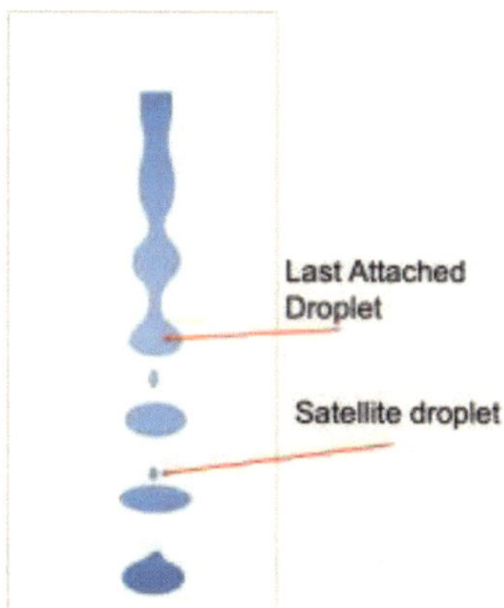

Figure 3.
The stream breaks up at a rate that is determined by the frequency of the piezo vibration. The last attached droplet is very important, as this is the point at which the computer determines exactly when to put a charge on the stream. As the droplet leaves the stream, it carries the same charge as was placed on the stream for that droplet. Thus, this droplet can then be manipulated to one side or other of the deposition area under the stream.

The sort stream itself has some important components that must be appreciated in order to understand how the sorting mechanism operates. The stream is generally a salt solution that can carry a charge. As can be seen in **Figure 3**, when the piezo crystal is activated the stream breaks up into droplets at a rate determined by the frequency of vibration. While these droplets can contain the particle of interest, there are also sub-droplets, termed satellite droplets, that break away from the main drops, as noted in **Figure 3**. As a rule, the higher the frequency of vibration, the smaller the droplet size and consequently the smaller the particle or cell that can be sorted. Other factors to be considered include the sorting pressure, which can be very high and can cause cellular damage.

3. Fluorescence-based cell sorting

The development of single-cell sorting as a new technology had little impact until the integration of fluorescence detection and laser-based excitation. These achievements emerged from the laboratory of Len and Lee Herzenberg at Stanford University as the process of fluorescence-based cell sorting that they termed fluorescence-activated cell sorting, or "FACS." This term was subsequently trademarked by Becton Dickinson, who first commercialized the Herzenberg technology. The early work from this laboratory used an intracellular fluorochrome (fluorescein diacetate) to label cells with very bright fluorescence that was excited by a mercury-arc light source; signals were detected by a single photomultiplier tube (PMT) [4]. Thus began decades of development from this laboratory: first an integrated laser excitation source [3], followed by multiple lasers and scatter detectors [17], advanced light-scatter analysis [18], and ultimately the first multiple fluorescence detection using two PMTs [8], the last of which (as we know) expanded over the next three decades to the current demonstration of 40 colors [19]. The Herzenbergs were without doubt the most significant team in the field of flow-cytometry cell sorting, and their developments over four decades created the foundation for an entire field of fluorescence-based cell sorting as well as core applications of fluorescence-based analysis. Len Herzenberg was awarded the Kyoto Prize in 2006 in honor of his accomplishments.

4. Magnetic particle sorting

Cell sorting using magnetic beads is a fast and efficient technology for separating cells without the need for the sort of high technology incorporated into flow cytometry. The huge advantage is the speed of separation and the fact that very large volumes of sample can be processed. The technology was developed by Miltenyi [20] in 1990 and is well embedded within current separation techniques today. In quick succession, the technology was used to separate pituitary cells [21], endothelial cells [22], B cells [23], eosinophils [24], proviral DNA [25], CD34+ cells [26], T cells [27], and bacteria [28]. The use of magnetic beads for sorting is demonstrated in **Figure 4**: a typical setup shows how magnetic beads attached to

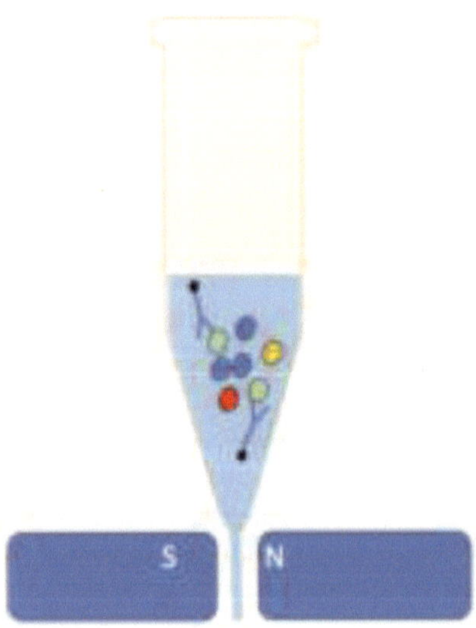

Figure 4.
A typical setup for magnetic sorting. Samples in the upper sample container can be separated through the addition of magnetic beads, typically labeled with an antibody to the target cell type. When the samples are passed by a powerful magnet, the magnetic beads and their targets are retained and thus purified. This technique can also be applied in large culture dishes such as 96- or 384-well plates.

antibodies are able to selectively separate cells or particles, as these are captured by the magnetic field and remain in the container while all other cells or particles unattached to the beads are removed.

The advantages of this technology are the very high speed of separation and the applicability to large volumes of sample. The disadvantage is that the selected population is still attached to a bead and an additional operation is required to separate the bead from the cell. However, this technique is very effective in isolation of rare cells when an appropriate marker is available. Another advantage of magnetic-bead separation is that if the beads are of different size or can be marked with dyes of different intensities, it is then possible to perform multiplexed assays to target multiple analytes simultaneously. The disadvantage of this technique is the fact that you either use negative selection, i.e., you target everything except the phenotype you want, or must use a magnetic bead attached to your selected phenotype.

5. Microfluidic chip–based sorting

The fundamental principle of microfluidic sorting is the use of narrow-bore tubes through which particles flow. Depending on the type of chip, it is possible to use a variety of methods to move cells of interest to flow toward one output or another, as shown in **Figure 5**. With the advent of new manufacturing technology over the past couple of decades, many systems utilize microfluidic chip–based sorting effectively. The advantage is that very small volumes, even less than a microliter, can be processed in complex lab-on-chip environments in which reaction sequences can be integrated within the chips. Many types of microfluidic chips have been developed that can physically separate cells, particles, and droplets at very high rates, making these lab-on-a-chip approaches very popular.

6. The limits of cell sorting

Cell-sorting technology is by no means a single entity or technology. There are several commercial instruments on the market, each with its advantages and disadvantages. However, some basic elements of sorting are fundamental to all instruments. For example, the goal of sorting differs depending on the sample being sorted. A sample containing desired rare cells to be isolated creates a significant problem that must be approached differently from a sample containing 20% of the desired phenotype. In the latter case, the sorting speed may not be a

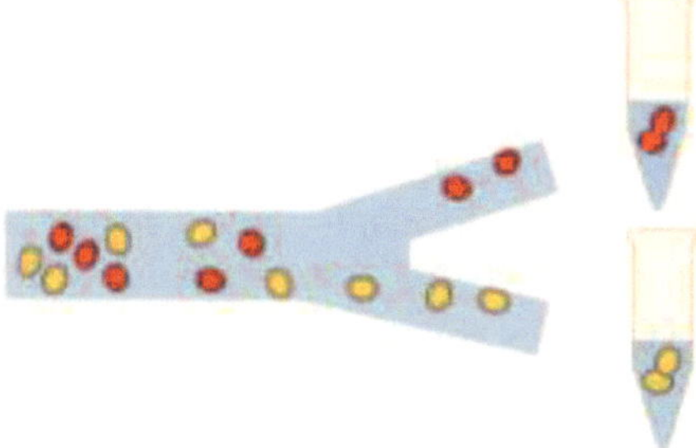

Figure 5.
A generic microfluidic channel with 2 outputs. The process for selecting one or another population depends on the nature of the chip. Several different techniques have been demonstrated.

significant issue at all. However, if the sample contains 0.01% or less of desired cells, then many tens or hundreds of thousands of cells must be evaluated before even a few desired cells can be collected. It has not been uncommon for cell sorters to be running 10-15 hours to obtain sufficient cells of rare populations in order to perform a desired experiment. Instrument designers were naturally driven to create instruments that could operate at far higher sampling rates than previously; an entire generation of high-speed sorters was designed, starting with the MoFlo that emerged in the late 1990s and was without doubt the most significant commercial sorter available.

Sorters can be programmed to increase yields, for example where maximum recovery of desired cells is established. This may mean that undesirable cells contaminate the sorted sample, but the operator is prepared to accept this in return for maximum recovery. An alternative is sorting for purity, where no undesirable cells are acceptable. However, in this mode, it is likely that the operator will lose some desirable cells.

Another consideration has been instrument dead-time – the time between calculations of the sorting algorithm during which the instrument cannot make a decision. In such cases, cells will be lost. Cell sorting is a complex process that ultimately depends upon the frequency of coincident events that occur at any stage of the sorting process. The process is determined by Poisson statistics ultimately based on the probability of observing a fixed number of events within a fixed time where we know the probability of an event's occurring [29]. We can thus define the probability of a particle's being in a droplet or being lost to coincidence, for example. Such calculations in essence determine the efficiency of instrument sorting capacity. Many factors impact these decisions, such as the type of sort (purity or recovery) that one desires and the limits that one allows for errors.

The maximum sort speed is based on several factors, one of which is the electronics cycle time. Given that normal analysis time is in the range of 3-5 μs, this translates to a range around 250,000 events per second. However, achieving these rates would require enormous pressures that would be counterproductive to the use of most cellular materials, so the sample pressure may well be one of the limiting factors in cell sorting.

The fundamental process for droplet formation was defined over 150 years ago by Lord Rayleigh [30] in his treatise on acoustics. Essentially, based on Rayleigh criteria, a cylinder of liquid with a diameter D will break into drops spaced by $\lambda = \pi D$. While the goal of this chapter is not to discuss the fluid dynamics in detail, it is sufficient to note that the relationship between nozzle diameter, pressure, and frequency defines the characteristics of a particular system. These details are well described by van den Engh [31].

7. The advantages of cell sorting

While numerous techniques exist for establishing bulk cell populations, based on viscosity or by using magnetic beads, for example, there are few methods whereby absolute purity can be created from mixed populations, and even fewer where this can be achieved at the single-cell level. Cell sorting in flow cytometry can be achieved by use of a targeting fluorescent-conjugated antibody tag to extract the targeted cell type. The primary advantage of flow-cytometry sorting is clearly the ability to separate individual cells from complex mixtures given that one has access to appropriate targeting antibodies. Cell sorting is the most efficient method currently available for creating single-cell populations from complex mixtures. One advantage of flow cytometry is that this process can be achieved without damaging the cells to any significant extent, in either a normal or a sterile environment. The ability of flow cytometry to sort under sterile conditions allows the sorting of cells of interest for cloning, frequently into 96-well plates for subcloning cells of interest [32–34].

Some cell-sorting applications do not need complexity of signals, but instead accuracy and purity based on very few parameters. One such application is sperm sorting, which has become an area of significant economic impact in flow cytometry. From the earliest concepts of evaluating X and Y chromosome–bearing sperm from domestic animals [35] to the effective use of cell sorters to electronically separate animal male and female sperm based on DNA content [11, 36, 37] for subsequent sperm sorting, flow cytometry has been the only really successful technology. Indeed, flow cytometry has also been extensively used for sorting human sperm [38], with extensive studies showing that the presence of Hoechst dye combined with UV excitation does not damage human sperm [39]. This approach was extended to sorting human sperm for in-vitro fertilization [40], a procedure that was allowed for several years by the Federal Drug Administration (FDA) but was eventually withdrawn, mostly for ethical reasons, a few years ago. What has persisted, however, with regard to sperm analysis, is the well-defined analytical assays for determination of sperm quality [41–44].

A clear advantage of flow cytometry–based cell sorting is the ability to separate a unique signature in a highly heterogeneous environment. With significant advances in recent years in availability of and options in monoclonal antibodies, almost every cell type possible can be targeted. The vast increase in the number of available fluorochromes has enabled the near effortless separation of complex mixtures.

8. Next-generation cell sorting

To move to the next generation of cell sorting, an instrument must have capacity for multiple lasers, a large number of detectors, and ability to sort under multiple algorithmic processes. Ideally a next-generation sorter has the ability to utilize spectral unmixing as well. The principle of spectral flow cytometry was developed some years ago by our laboratory [45–49]; however, it was restricted to analysis instruments. More recently, a commercially available instrument called "Bigfoot," a sophisticated and advanced sorter developed by Propel and now owned by ThermoFisher, is capable of real-time spectral unmixing for sort-decision making, which will be discussed later.

Over the past several years several companies have developed sorters with impressive automation capabilities. One example is the Sony SH800 cell sorter. In developing this sorter, Sony clearly identified one of the classic failures of sorter manufactures – lack of automation of instrument setup and calibration. As stated in the instrument brochure, "All setup steps, including optical alignment, droplet formation, side stream calibration, and delay time adjustment, are automated using Sony-developed CoreFinder technology™. This completely eliminates all the complicated setup work required with conventional cell sorters." This implementation by Sony is crucial; previous sorter manufacturers had to a large extent ignored this task because they assumed that all their sorters required highly skilled technologists to perform critical alignments. Earlier instruments had established components that assisted calibration and setup. For example, in the early 1990s, Coulter Corporation (now Beckman Coulter (BC)) integrated a video camera and screen to view the droplet breakoff. Interestingly, the Coulter system showed a screen with a horizontal stream image. Becton Dickinson (BD) also installed a camera in the next iteration of *their* sorter but made the screen vertical in the same path as the operator would view the stream using the previous high-technology approach (a telescope). Such iterations perpetuated iterative updates with useful but not really transformational features for each version of sorter instrument. Sony therefore broke the mold and moved to a fully automated setup that was faster and more accurate than even a highly trained technician could achieve.

Another important development implemented by Sony was the use of a disposable sorting "chamber." It was in fact a sorting chip and is shown in **Figure 6**. Integrating a disposable chip into a cell sorter required many innovative modifications to the sorting design that even changed some time-honored expectations of how a sorter should operate. The integration of Sony's outstanding mechanics in the x-y-z stage for plate sorting was also innovative in implementation. This allowed for index sorting that gave the user confidence regarding the precision of sort decisions. While Sony had not been previously competing in the cell-sorter market, they focused heavily on features that promoted automation in setup and operation and as such drove competitive products to accommodate these features, resulting in significant advantages to the user base in the cytometry field.

Cell sorting can be achieved using most particles or cell types. As noted earlier, sorting can be achieved directly into test tubes, 96-well plates, or for that matter virtually any type of plate. In addition, we have used cell sorting for bacterial sorts directly onto petri dishes as a single bacterium or in multiples (**Figure** 7).

The recent debut of the top-level Bigfoot cell sorter again broke many previously accepted norms for sorting instruments. Apart from the name, which as a previous Everest summiter (May 23, 2009) I will happily ignore, the Bigfoot is an almost fully automated 60-channel, 9-laser sorter. One of the most pleasant features is almost silent operation. Researchers who sorted for decades on other instruments became used to noisy pressure and vacuum pumps running almost continuously. The Bigfoot is pleasantly silent. In addition, the sorter has what I would classify as the safest operating environment of all sorters. While integration of sorters within class-II safety cabinets began in the early 2000s and was copied across the industry, the Bigfoot chose a more effective route. The engineers integrated a custom-designed class-II safety chamber as part of the instrument itself; within this chamber is another chamber for the actual sorting. There is little doubt that this sorter can safely accommodate pathogens or human samples.

Bigfoot embodies multiple features that I consider entirely sensible and that will no doubt be copied across the industry. Remote control of the instrument is effective

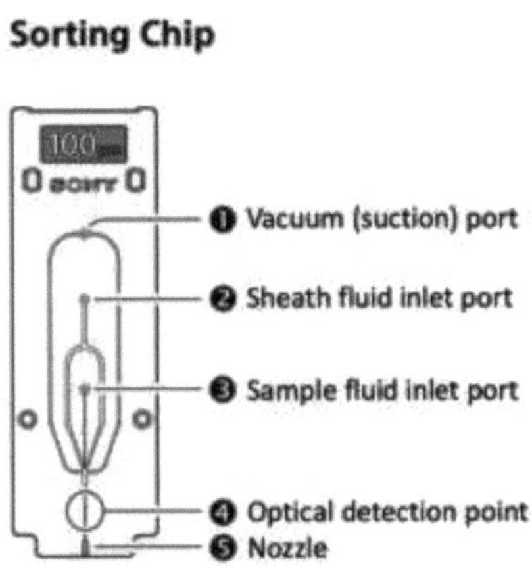

Figure 6.
Sony's innovative sorting chip introduced into their cell-sorter line was mass produced and could easily be replaced. (figure modified from Sony website).

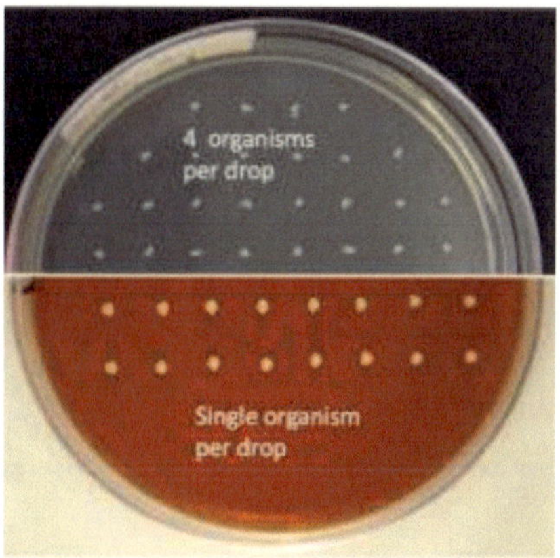

Figure 7.
Single organisms can be sorted directly onto petri dishes. On the top, organisms were sorted at four organisms per "well" location, and on the bottom a single organism was plated in a 96-well format.

and well managed, with direct support from company engineers available in minutes. The implementation of bead standards within a cooled vessel allows the instrument to proceed through setup and calibration without any user input except initiation of the software process. Combining this feature with remote control allows a user to literally prepare the instrument for sorting, even prior to entering the laboratory, a time-saving feature that I find valuable and practical. Similarly, the system comes standard with six sample stations that a user can fill and then proceed with virtually automated sorting without having to physically enter the sorting environment. For sorting of pathogenic samples, or samples with potential infectious agents, this feature is both innovative and a really sensible and practical advantage to the user.

The other major feature of this instrument is its capability for performing spectral unmixing and making sort decisions on those calculated curves. This is the first instrument capable of spectral sorting and it opens up an entirely new area of usage owing to the increase in power of spectral flow cytometry. In the years since it was first developed by our group [45, 46, 50], spectral flow cytometry has come to dominate the interest of immunologists and cytometry-related scientists. One reason is that spectral cytometry captures all the information available from multiple detectors and can resolve spectra that are very similar, but that emanate from different chemical species [51]. This discrimination allows a greater range of chromophores to be integrated into a highly complex panel. An example of this is the recent demonstration of 40-color phenotyping [19]. For decades there has been a rivalry to increase the number of simultaneous fluorophores in flow cytometry, which began with a single fluorophore and now is at 40. The impact of spectral cytometry is certain to pay dividends as the need to physically sort spectrally defined populations becomes evident.

While spectral analysis has been implemented for the past several years, the ability to sort based on a specific spectrally defined population has not been possible. With the implementation of Bigfoot's spectral sorting capacity (see **Figure 8**), physical sorting based on real-time spectral-unmixing algorithms at rates approaching 70,000 cells/second is now possible.

Flow cytometry sorting has changed significantly in recent years after many decades of iterative changes. In the 1980s and early 1990s, almost all users were focused on cell sorters as the core support instruments in their laboratories. This changed with the rapid grown of small analyzers, which became the primary instrument for flow cytometry, being efficient as well as cheaper and easier to manage than cell sorters. Ironically, the growth of the analyzer market and the excellence of competitive technologies resulted in a resurgence of demand for cell sorters with highly advanced features.

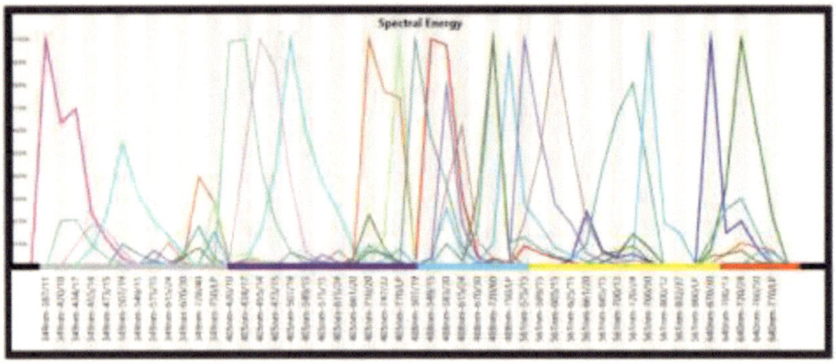

Figure 8.
An example of Bigfoot's spectral capability, which can be implemented within its sorting algorithms to perform sorting based on spectral separation.

9. Summary

Cell-sorting technologies have advanced over the 55 years since single-cell separation techniques were first invented. Of the multiple tools in the cell-separation toolkit, the most effective so far has been flow cytometry based on fluorescence detection. The combination of multiparameter scatter signals from laser excitation and fluorescence emission from multiple detectors provides a unique multiparameter analysis capability, which together with sorting algorithms gives flow-cytometry sorters unique capabilities in separating even very rare cell populations with excellent purity. In the first two decades of the emergence of flow cytometry, cell sorting was the predominant implementation. However, within the last couple of decades the predominant impact in the field has been analytical instruments, not sorters. In fact, these analytical instruments have become so advanced that they have expanded the field significantly as very high levels of complexity have driven tremendous immunological advances. What has resulted from this is a new demand for cell sorters with equal levels of complexity, with the unfortunate large increase in cost, but the advantage of vastly increase automation. If there are any drawbacks to this direction, it is the cost of instruments that can now exceed $1 M each. These instruments are most likely going to be in core laboratories that service multiple clients, since the average laboratory or even small companies cannot afford this level of expenditure. Regardless of these issues, piezo-based sorting is still the fastest and most efficient technique available for the sorting of any complex mixture.

Integration of a variety of detection approaches has enable a vast range of microfluidic systems – typically lab-on-a-chip technologies. These lab-on-chip approaches are heavily focused on diagnostic assays, which are likely to become a much larger segment of the cell-sorting market. The emergence of spectral unmixing using multi-array detectors has expanded both detection and sorting technology, largely owing to the high-speed FPGA-based electronics necessary for the current high-complexity sorting instruments.

Acknowledgements

The author is grateful to support from the 1R56AI089511-01, R21RR033593-01 & 1R33CA140084 and 3OT2OD023847-01S5. NSF Grant#2027049-CBET. U.S. Department of Agriculture, project 1935-42000-072-02G, Center for Food

Safety Engineering at Purdue University. NIFA Grant # 2020-70410-32897. Thanks to Ms. Kathryn Ragheb and Ms. Brianna Dowden for flow-cytometry support.

Conflict of interest

The author declares no conflict of interest.

Author details

J. Paul Robinson
Purdue University, West Lafayette, IN, USA

*Address all correspondence to: wombat@purdue.edu

References

[1] Fulwyler MJ. Electronic separation of biological cells by volume. Science 1965;150:910-911.

[2] Lushbaugh CC, Hale DB. Electronic measurement of cellular Volume. VII. Biologic evidence for two volumetrically distinct subpopulations of red blood cells. *Annual Report Biological and MedicalResearch Group (H-4) of the Health Division, July 1962 through June 1963.* Los Alamos Scientific Laboratory of the University of California; 1963:270-278.

[3] Bonner WA, Hulett HR, Sweet RG, Herzenberg LA. Fluorescence activated cell sorting. Review of Scientific Instruments 1972;43(3):404-409.

[4] Hulett HR, Bonner WA, Barrett J, Herzenberg LA. Cell sorting: automated separation of mammalian cells as a function of intracellular fluorescence. Science 1969;166(3906):747-749.

[5] Herzenberg LA, Black SJ, Loken MR, Okumura K, van der Loo W, Osborne BA, et al. Surface markers and functional relationships of cells involved in murine B-lymphocyte differentiation. Cold Spring Harbor Symposia on Quantitative Biology 1977;41 Pt 1:33-45.

[6] Herzenberg LA, Sweet RG, Herzenberg L. Fluorescence-activated cell sorting. Scientific American 1976;234:108-117.

[7] Loken MR, Parks DR, Herzenberg LA. Identification of cell asymmetry and orientation by light scattering. Journal of Histochemistry and Cytochemistry 1977;25:790-795.

[8] Loken MR, Parks DR, Herzenberg LA. Two-color immunofluorescence using a fluorescence-activated cell sorter. Journal of Histochemistry and Cytochemistry 1977;25:899-907.

[9] Gledhill BL, Lake S, Steinmetz LL, Gray JW, Crawford JR, Dean PN, et al. Flow microfluorometric analysis of sperm DNA content: effect of cell shape on the fluorescence distribution. Journal of Cellular Physiology 1976;87:367-375.

[10] Evenson DP, Darzynkiewicz Z, Melamed MR. Relation of mammalian sperm chromatin heterogeneity to fertility. Science 1980;210:1131-1133.

[11] Johnson LA, Flook JP, Look MV. Flow cytometry of X and Y chromosome-bearing sperm for DNA using an improved preparation method and staining with Hoechst 33342. Gamete Research 1987;17(3):203-212.

[12] Arndt-Jovin DJ, Jovin TM. Analysis and sorting of living cells according to deoxyribonucleic acid content. Journal of Histochemistry and Cytochemistry 1977;25:585-589.

[13] Carrano AV, Gray JW, Langlois RG, Burkhart-Schultz KJ, Van Dilla MA. Measurement and purification of human chromosomes by flow cytometry and sorting. Proceedings of the National Academy of Sciences 1979;76:1382-1384.

[14] Green DK, Malloy P, Steel M. The recovery of living cells by flow sorting machine. Acta Pathologica et Microbiologica Scandinavica Supplementum 1981;274:103-106.

[15] Visser JWM, Bol SJL, van den Engh G. Characterization and enrichment of murine hemopoietic stem cells by fluorescence activated cell sorting. Experimental Hematology 1981;9:644-655.

[16] Dangl JL, Parks DR, Oi VT, Herzenberg LA. Rapid isolation of cloned isotype switch variants using fluorescence activated cell sorting. Cytometry 1982;2(6):395-401.

[17] Hulett HR, Bonner WA, Sweet RG, Herzenberg LA. Development and application of a rapid cell sorter. Clinical Chemistry 1973;19(8):813-816.

[18] Loken MR, Sweet RG, Herzenberg LA. Cell discrimination by multiangle light scattering. Journal of Histochemistry and Cytochemistry 1976;24:284-291.

[19] Park LM, Lannigan J, Jaimes MC. OMIP-069: Forty-color full spectrum flow cytometry panel for deep Immunophenotyping of major cell subsets in human peripheral blood. Cytometry Part A 2020;n/a(n/a).

[20] Miltenyi S, MÅller W, Weichel W, Radbruch A. High gradient magnetic cell separation with MACS. Cytometry 1990;11:231-238.

[21] Wynick D, Bloom SR. Magnetic bead separation of anterior pituitary cells. Neuroendocrinology 1990; 52:560-565.

[22] Hewett PW, Murray JC. Human microvessel endothelial cells: isolation, culture and characterization. In Vitro Cell Dev.Biol.Anim 1993;29A(11): 823-830.

[23] McKnight ME, Mukerjee S, Nasoff M, Gaskins C, Glassy MC. Isotype-directed enrichment of B cells by magnetic beads in the generation of immunoreactive human monoclonal antibodies. Hybridoma 1996;15(4): 255-261.

[24] Sedgwick JB, Shikama Y, Nagata M, Brener K, Busse WW. Effect of isolation protocol on eosinophil function: Percoll gradients versus immunomagnetic beads. J.Immunol.Methods 1996;198(1): 15-24.

[25] Dorenbaum A, Venkateswaran KS, Yang G, Comeau AM, Wara D, Vyas GN. Transmission of HIV-1 in infants born to seropositive mothers: PCR-amplified proviral DNA detected by flow cytometric analysis of immunoreactive beads. J.Acquir.Immune.Defic.Syndr. Hum.Retrovirol. 1997;15(1):35-42.

[26] Kawano Y, Takaue Y, Law P, Watanabe T, Abe T, Okamoto Y, et al. Clinically applicable bulk isolation of blood CD34+ cells for autografting in children. Bone Marrow Transplant. 1998;22(10):1011-1017.

[27] Maeda S, Miyawaki T, Nakanishi T, Takigawa M, Shimada M. Peripheral type benzodiazepine receptor in T lymphocyte rich preparation. Life Sciences 1998;63(16):1423-1430.

[28] Porter J, Robinson J, Pickup R, Edwards C. An evaluation of lectin-mediated magnetic bead cell sorting for the targeted separation of enteric bacteria. J.Appl.Microbiol. 1998;84: 722-732.

[29] Taylor JR. An introduction to Error Analysis. Sausilito, CA: University Science Books; 10997.

[30] Raleigh JWS. Theory of Sound. London, Reprinted 1945, Dover, New York: mCMillan; 1877.

[31] van den Engh G. High Speed Cell Sorting. In: Durack DaR, J. P., editor Emerging Tools for SIngle Cell Analysis. New York: Wiley-Liss; 2000, p. 2148.

[32] Lei L, Tran K, Wang Y, Steinhardt JJ, Xiao Y, Chiang C-I, et al. Antigen-Specific Single B Cell Sorting and Monoclonal Antibody Cloning in Guinea Pigs. Frontiers in microbiology 2019;10:672.

[33] Smith MJ, Packard TA, O'Neill SK, Hinman RM, Rihanek M, Gottlieb PA, et al. Detection and Enrichment of Rare Antigen-specific B Cells for Analysis of Phenotype and Function. Journal of Visualized Experiments : JoVE 2017(120):55382.

[34] Zhou Y, Liu Z, Wang Z, Zhang Q, Mayer CT, Schoofs T, et al. Single-Cell Sorting of HBsAg-Binding Memory B Cells from Human Peripheral Blood Mononuclear Cells and Antibody Cloning. STAR Protocols 2020;1(3): 100129.

[35] Garner DL, Gledhill BL, Pinkel D, Lake S, Stephenson D, Van Dilla MA, et al. Quantification of the X- and Y-chromosome-bearing spermatozoa of domestic animals by flow cytometry. Biology of Reproduction 1983;28: 312-321.

[36] Johnson LA. Separation of X and Y chromosome-bearing sperm on the basis of flow cytometric analysis for DNA. No Journal Name 1992.

[37] Johnson LA. Sex preselection by flow cytometric separation of X and Y chromosome-bearing sperm based on DNA difference: a review. Reprod.Fertil. Dev. 1995;7(4):893-903.

[38] Auger J, Leonce S, Jouannet P, Ronot X. Flow cytometric sorting of living, highly motile human spermatozoa based on evaluation of their mitochondrial activity. Journal of Histochemistry and Cytochemistry 1993;41(8):1247-1251.

[39] Catt SL, Sakkas D, Bizzaro D, Bianchi PG, Maxwell WM, Evans G. Hoechst staining and exposure to UV laser during flow cytometric sorting does not affect the frequency of detected endogenous DNA nicks in abnormal and normal human spermatozoa. Mol.Hum.Reprod. 1997;3(9):821-825.

[40] Vidal F, Fugger EF, Blanco J, Keyvanfar K, Catala V, Norton M, et al. Efficiency of MicroSort flow cytometry for producing sperm populations enriched in X- or Y-chromosome haplotypes: a blind trial assessed by double and triple colour fluorescent in-situ hybridization. Human Reproduction 1998;13(2):308-312.

[41] Aravindan GR, Bjordahl J, Jost LK, Evenson DP. Susceptibility of human sperm to in situ DNA denaturation is strongly correlated with DNA strand breaks identified by single-cell electrophoresis. Experimental Cell Research 1997;236(1):231-237.

[42] Evenson DP, Darzynkiewicz Z, Melamed MR. Comparison of human and mouse sperm chromatin structure by flow cytometry. Chromosoma 1980;78:225-238.

[43] Evenson DP, Darzynkiewicz Z, Melamed MR. Simultaneous measurement by flow cytometry of sperm cell viability and mitochondrial membrane potential related to cell motility. Journal of Histochemistry and Cytochemistry 1982;30:279-280.

[44] Spano M, Evenson DP. Flow cytometric analysis for reproductive biology. Biology of the Cell 1993;78(1-2):53-62.

[45] Grégori G, Patsekin V, Rajwa B, Jones J, Ragheb K, Holdman C, et al. Hyperspectral cytometry at the single-cell level using a 32-channel photodetector. Cytometry Part A 2012;81A(1):35-44.

[46] Robinson JP. Multispectral cytometry: The next generation. Biophotonics International 2004;October, 2004 p36-40.

[47] Robinson JP, Grégori G, Rajwa B, Jones J, Patsekin V. Multispectral detector and analysis system. In: USPTO, ed. USA: Purdue University; 2007.

[48] Robinson JP, Patsekin V, Grégori G, Rajwa B, Jones J. Multispectral flow cytometry: Next generation tools for automated classification. Microsc. Microanal. 2005;11(S02):2-3.

[49] Robinson JP, Rajwa B, Grégori G, Jones J, Patsekin V. Multispectral cytometry of single bio-particles using a 32-channel detector. In: Dinh TV, Grundfest WS, Benaron DA, Cohn GE, eds. 5692. Bellingham, WA: SPIE; 2005:359-65.

[50] Robinson JP, Rajwa B, Grégori G, Jones J, Patsekin V. Collection hardware for high speed multispectral single particle analysis Cytometry 2004;59A:12-.

[51] Robinson JP. Spectral flow cytometry—Quo vadimus? Cytometry Part A 2019;95(8):823-824.